Social Organization
of the Rufous Vanga

Social Organization of the Rufous Vanga

The Ecology of Vangas—Birds Endemic to Madagascar

Satoshi YAMAGISHI
ed.

First published in 2005 jointly by:

Kyoto University Press
Kyodai Kaikan
15-9 Yoshida Kawara-cho
Sakyo-ku, Kyoto 606-8305, Japan
Telephone: +81-75-761-6182
Fax: +81-75-761-6190
Email: sales@kyoto-up.gr.jp
Web: http://www.kyoto-up.gr.jp

Trans Pacific Press
PO Box 120, Rosanna, Melbourne
Victoria 3084, Australia
Telephone: +61 3 9459 3021
Fax: +61 3 9457 5923
Email: info@transpacificpress.com
Web: http://www.transpacificpress.com

Set by KWIX Co., Ltd.

Printed in Nagoya (Japan) by KWIX Co., Ltd.

Distributors

Australia
Bushbooks
PO Box 1958, Gosford, NSW 2250
Telephone: (02) 4323-3274
Fax: (02) 9212-2468
Email: bushbook@ozemail.com.au

UK and Europe
Asian Studies Book Services
Franseweg 55B, 3921 DE Elst
Utrecht, The Netherlands
Telephone: +31 318 470 030
Fax: +31 318 470 07349
Email: info@asianstudiesbooks.com
Web: http://www.asianstudiesbooks.com

USA and Canada
International Specialized Book
Services (ISBS)
920 NE 58th Avenue, Suite 300
Portland, Oregon 97213-3786
USA
Telephone: (800) 944-6190
Fax: (503) 280-8832
Email: orders@isbs.com
Web: http://www.isbs.com

ISBN 1 920901 04 3

National Library of Australia Cataloging in Publication Data

The Ecology of Vangas–Birds Endemic to Madagascar

Bibliography.
Includes index.
ISBN 1 920901 04 3.

Social Organization of the Rufous Vanga

Contents

PART 1

*

The Ampijoroa Field Station

PART 2

*

Social organization of the Rufous Vanga

PART 3

*

Tracking the route taken by Rufous Vangas

Credit for Figures and Illustrations

Shigeki Asai Plate I-9, Plate I-11, Plate I-14, Plate II-2, Fig 5-1

Conservation International Plate I-4

Keiko Kanao The title page of Part III, Plate III-1

Tomohisa Masuda Cover photo, Plate I-6, Plate I-8, Plate I-12 (b, c, d, f), Plate I-15, Plate I-15, Plate II-5

Akira Mori Fig 2-13

Masahiko Nakamura Plate I-10, Plate I-12 (a), Fig 8-1

Yasuko Segawa Plate I-13

Eiichiro Urano The title page of Part II, Fig 5-3

Masaru Wada Plate II-6

Satoshi Yamagishi The title page of Part I, Plate I-1, Plate I-2, Plate I-12 (e), Plate II-1, Plate II-3, Plate II-4, Plate II-7, Fig 2-2, Fig 2-3, Fig 2-9, Fig 2-10, Fig 2-11, Fig 2-12, Fig 2-14, Fig 3-2, Fig 3-3, Fig 3-4, Fig 3-5, Fig 5-2, Fig 5-5, Fig 8-2, Fig 8-3

The Ampijoroa Field Station

Children of Ampijoroa Village at the entrance of the forest.

a

b

c

Plate I-1 Three Rufous Vanga individuals which visit to the same nest.

a: young male (the individual banded with pink stripes)

b: adult male

c: adult female

Plate I-2 Morning mist rises over the fresh green of Ankarafantsika National Park.

Plate I-3 Elen, a German graduate student, shoulders a boa.

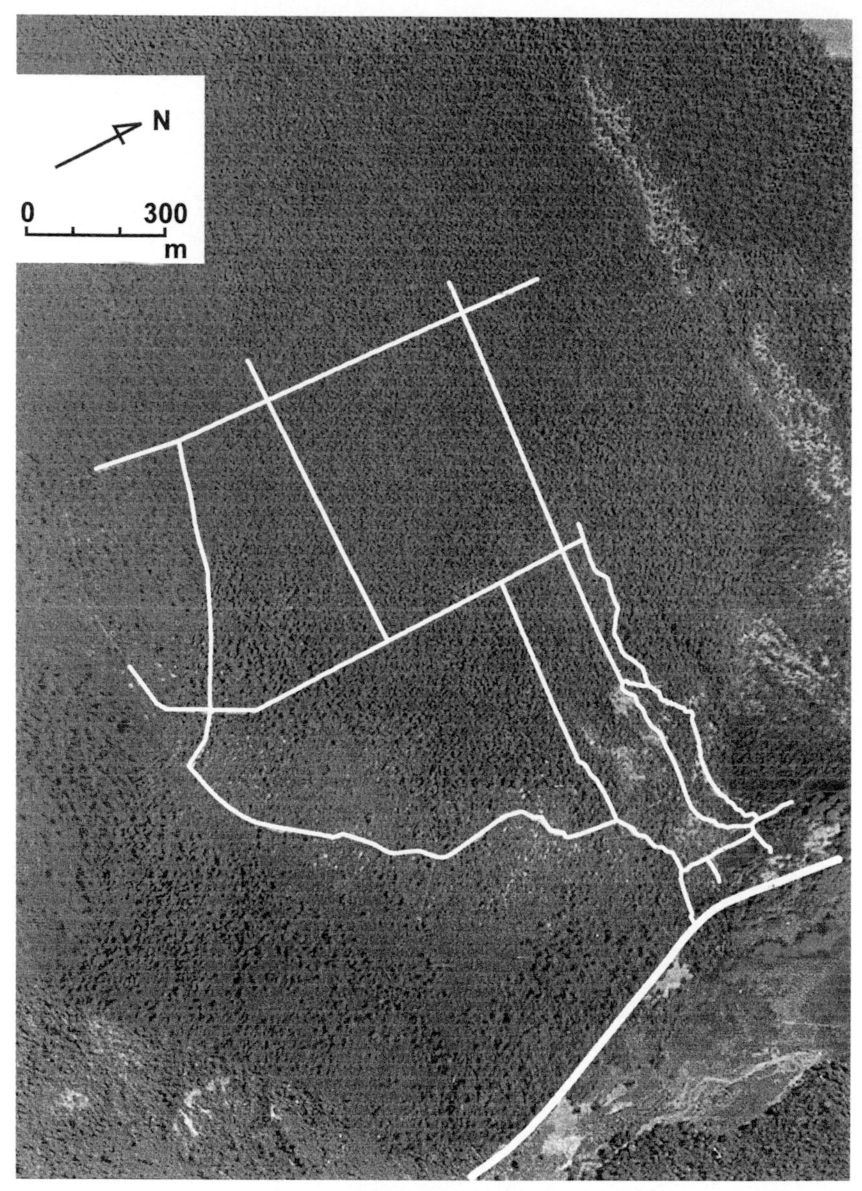

Plate I-4 An aerial photograph of Jardin A research field. The main routes correspond those shown in Fig. 2-4 (photo provided by Conservation International).

Plate I-5 Inside the research field in a deciduous broad-leaved forest.

Plate I-6 Nile crocodile (*Crocodylus niloticus*) basking on the lakeshore of Lake Ravelobe.

Plate I-7 Villagers washing and fishing on the lakeshore of Lake Ravelobe. They fish with a line, without using a rod.

Plate I-8 Some spotted-throat males (yearlings) have fairly blackish throat.

Plate I-9 A female (right) nestles up to a male (left).

Plate I-10 Van Dam's Vanga breeding in Jardin A.

a

b

c

Plate I-11 The head and throat area of the Rufous Vanga.
a: yearling male (with spots)
b: two-year-old and over (adult) male
c: female
(from Yamagishi et al. 2002)

Plate I-12 Main characters in the mixed-species flock theatre company.
a: Blue Vanga b: Crested Drongo c: Paradise Flycatcher d: Common Newtonia
e: Long-billed Greenbul f: Ashy Cuckoo-Shrike

Plate I-13 Seek shelters within a flock of Rufous Vanga.
In their breeding season the Rufous Vanga emits an alarm call immediately after their natural enemy, such as raptors and lemurs, appears, and occasionally they even daringly chase around the predators en masse. It is considered that other members in the mixed species flock one-sidedly tag along with Rufous Vangas is because the defensive prowess of the Rufous Vanga is attractive to them. (Drawn by Y.Segawa)

Plate I-14 The buff color tip of the greater coverts (indicated with arrow) signifies a yearling.

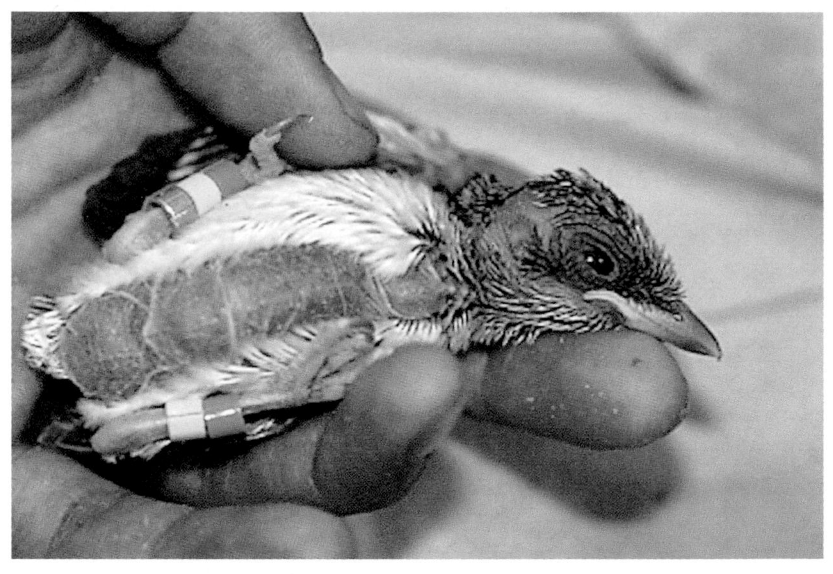

Plate I-15 Chicks are individually identified, being leg-banded with colored rings about six days after they hatched.

The motivation of the study

Satoshi YAMAGISHI

According to the *Dictionary of Japanese Names of Birds of the World*[1], the common Japanese name for the Rufous Vanga is Akaoohashimozu. From this phonetic notation, the non Japanese-speaking reader would scarcely surmise where to put pauses and inflections in pronunciation or imagine what the name indicates. A literal translation of the Chinese characters is the "large-beaked red shrike." The suitability of this name will be investigated later, in Chapter 8. Nevertheless, the biology of the Rufous Vanga, a bird endemic to Madagascar, is the main subject of this book.

From October 1991 to February 1992, Eiichirou Urano (then a research student of Osaka Municipal University) and I stayed at Ampijoroa (Plate I-2), approximately 100 km southeast of the port town of Mahajanga in northwest Madagascar (Fig 2-1). The objective of our stay was to conduct scientific research* on the adaptive radiation of a group of birds known to biologists as the family Vangidae. Concern over the prevailing political uncertainty in Madagascar led the Japanese Ministry of Foreign Affairs to advise researchers to voluntarily refrain from traveling there. However, the final decision on this matter was left to the discretion of the respective universities. Accordingly, Kazuhiro Eguchi of the Kyushu University, a member of our research team, had no option but to comply with his university's decision to follow this advice. He stayed behind in Japan, extremely disappointed and frustrated.

One day, while observing a nest of the Rufous Vanga *Schetba rufa*, we witnessed an unusual event. The breeding male (Plate I-1b) and female (Plate I-1c) were brooding the chicks after bringing food and feeding them when, to our great surprise, another bird appeared bearing food (Plate I-1a). At first, we could

* The study was supported by a Grant-in-Aid for International Scientific Research from the Japanese Ministry of Education, Culture, Sports, Science and Technology, No.01041079: Project head, Satoshi Yamagishi.

scarcely believe our eyes because more than 90% of the approximately 9,000 species of birds in the world breed monogamously, i. e., only the parents nurture the young. The bird in question was slightly smaller in size than the adult male and female. It had black spots extending from the neck to the throat, whereas an adult male has a black apron, and an adult female has a white breast.

The spotted-throat individual was provisioning the chicks and also participating in territorial defense, anti-predator defense, and, infrequently, in nest building. There were only two tasks that the spotted-throat individual did not perform (rather, it was apparently not allowed to perform): incubating the eggs and brooding the chicks. On the basis of these observations, we determined that the spotted-throat individual was a "home help"; in biology, this is more accurately termed a "helper." We then surmised that the spotted-throat individual might be a young male because its spotted region conformed to that of the black apron of an adult male, and its body size was marginally smaller than that of the adults.

The breeding system wherein breeding pairs are accompanied by auxiliary individuals (helpers) is termed "cooperative breeding," and this phenomenon had already been discovered over 50 years ago. In 1935[2], the American ornithologist A.F. Skutch contributed a paper entitled "Helpers at the nest" to The Auk (the journal of the American Ornithologists' Union). In this paper, Skutch reported that such breeding behavior was observed in three varieties of birds (the Central American Brown Jay *Psilorhinus mexicanus cyanogenys*, the Black-eared Bush-Tit *Psaltriparus melanotis melanotis*, and the Banded Cactus Wren *Heleodytes zonatus zonatus*). From his use of the word "helper" in the title of the paper, it is evident that he was convinced from the outset that the auxiliary individuals assist the parent birds. Cooperative breeding has been interpreted as a type of "altruistic behavior" since the activities of helpers benefit others but entail costs to themselves. We humans, scientists included, who are obsessed with (or who would like to be convinced of) the idea that "humans should behave altruistically," have drawn encouragement from the knowledge that "even birds behave altruistically."

Around 1965, "sociobiology" began to gain a great deal of attention. The factors underlying the evolution of social insect workers, an issue that tormented Darwin, were elucidated by Hamilton through the concept of inclusive fitness[3]. Subsequently, the evolution of helpers among birds, as an analogy of social insect workers, attracted the interest of numerous scientists. Today, more than half-a-century after Skutch's original observations, nearly 300 species of cooperatively breeding birds have been recognized. Adopting an investigative framework of kin selection, several studies on this subject have concluded that the activities of helpers serve to enhance the reproductive success of breeding pairs.

However, recent advances in the detailed study of this subject have shown that, in an increasing number of cases, the helper's aid does not enhance reproductive success to as great an extent as earlier believed. What about the case of the Rufous Vanga? During the year in which Urano and I made our surprise discovery, we undertook extensive preliminary research to investigate the Rufous Vanga's breeding system. Our research confirmed that eight out of 15 breeding pairs, i. e., more than half of those examined, were accompanied by more than one helper[4]. In the case of this bird, cooperative breeding is a common practice. Therefore, during the remainder of our stay, we caught as many Rufous Vangas as possible and banded each leg with a unique combination of colored rings that they were individually identifiable. Later, after our return to Japan, we eagerly awaited our next trip to Madagascar the following year in order to begin our main research. The bird we identified as helper No. 1 was banded on both legs with two pink-striped rings (Plate I-1a).

The project was implemented on the basis of our preliminary research, and it later developed into the main research project**, "Social Evolution of the Family Vangidae."

The project conducted detailed and exhaustive investigations of the breeding ecology and social system of the Rufous Vanga and also examined the ecology and behavior of other vangids and their taxonomic interrelationships. "Social Organization of the Rufous Vanga" has been authored by the researchers actively involved in the project. It describes for the first time, in an easy style, the methodology and implementation of the field study in Madagascar, and the novel and exciting findings of this research.

References

1 Yamashina, Y. (1986) *A World List of Birds with Japanese Names*. Daigaku Shorin, Tokyo (in Japanese).
2 Skutch, A. F. (1935) Helpers at the nest. *Auk,* 52: 257-273.
3 Hamilton, W. D. (1964) The genetical evolution of social behaviour. I. II. *Journal of Theoretical Biology*, 7: 1-52.
4 Yamagishi, S., Urano, E., & Eguchi, K. (1995) Group composition and contributions to breeding by Rufous Vangas *Schetba rufa* in Madagascar. *Ibis*, 137: 157-161.

**Supported by two grants from the Japanese Ministry of Education, Culture, Sports, Science and Technology Grant-in-Aid for the International Scientific Research Program, No.06041093: Research head, Satoshi Yamagishi, and Grant-in-Aid for Basic Research (A) (2), No.1169118: Research head, Satoshi Yamagishi.

Ampijoroa Forest, home to the Rufous Vanga

Taku Mizuta

The Ankarafantsika National Park and Ampijoroa village

After traveling approximately 450 km northwest from Antananarivo, the capital of Madagascar, on the National Road Route 4, we first crossed the bleak rolling plains of the central highlands and then the western arid savanna-like region before a thick green forest vista opened out before us. This was our destination-the Ankarafantsika National Park (Fig 2-1, Plate I-2). The National Park covers 60,520 hectares. The district wherein we conducted extensive research on birds, primarily on the Rufous Vanga *Schetba rufa*, is called the Ampijoroa Forest Station (20,000 hectares) and is located in the Ankarafantsika National Park.

The Ampijoroa Forest Station is adjacent to Route 4, which runs approximately through the center of Ankarafantsika. Until 1999, the forest station was administered and managed by an NGO named Conservation International (CI). This NGO was aided in its forest-management efforts, particularly the upgrading of the ecotourism program, by youth dispatched under the auspices of the U. S. Government's overseas volunteer Peace Corps program. After the year 2000, the administration of the Ampijoroa Forest Station passed from CI to the National Association for the Management of Protected Areas (Association Nationale pour la Gestion des Aires Protegees: ANGAP). In August 2002, Ankarafantsika attained the status of a national park under the management of ANGAP.

Other than ANGAP's administrative office, the forest station includes accommodation for researchers. The rooms are equipped with flush toilets and shower rooms (the showers spout cold water) (Fig 2-2). ANGAP's administrative office is also equipped with a solar power generation system, and the limited amount of electricity generated is used for office work and for

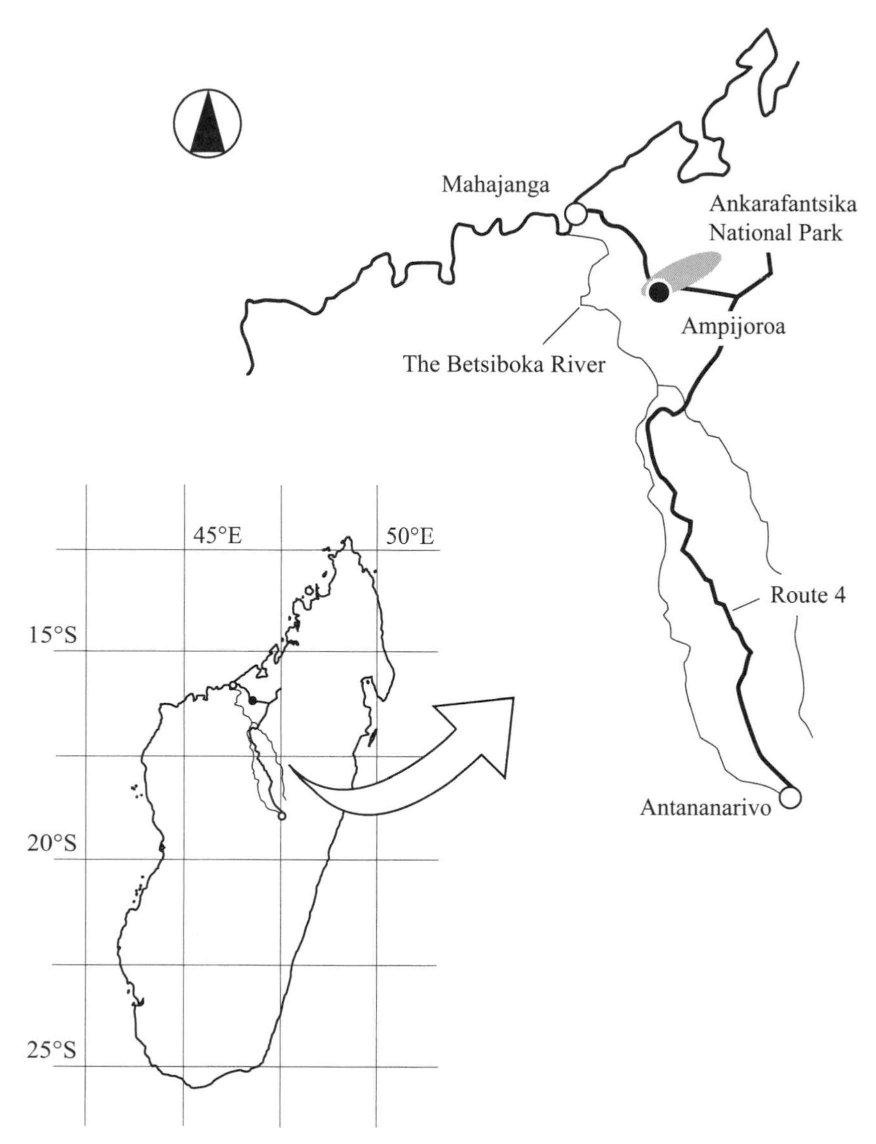

Fig 2.1 Map of Madagascar, and the route from the capitol Antananarivo to Ampijoroa Forest, location of our field study site.

Fig. 2-2 Accommodation for researchers.

providing light during the night. Thus, since the station is well-equipped with the necessary research facilities, scientists from all around the world reside there during study visits. During the six terms I spent at the station, I met several scientists from various other nations, such as Germany, Switzerland, France, Britain, Spain, the United States of America, and New Zealand (Plate I-3). The purpose of their visit was to undertake a variety of studies on a diverse array of plants and animals. Needless to say, the station also hosts several Malagasy researchers and graduate students. Thus, the ambience of the forest station is that of an international village of researchers.

The Ampijoroa forest in the Ankarafantsika National Park is also visited by several tourists, such as overseas natural history enthusiasts and tourist groups, on day trips or overnight tours arranged by a hotel approximately 100 km away from Mahajanga (Fig 2-1). Shortly after passing through the entrance facing Route 4, visitors arrive at the reception area where the entrance fees are collected (Fig 2-3). Accommodation and campsite facilities are available for tourists in order to ensure that their stay in Ampijoroa is comfortable. A small restaurant affords tourists the opportunity to enjoy homely Malagasy cuisine. Fairly "luxurious" grocery items, such as beer, soft drinks, cigarettes, and canned food, may be purchased from a small kiosk within the restaurant.

The ANGAP ecotourism venture offers tourists several sightseeing packages to explore the forest. Walking trails, along which visitors are guided by members

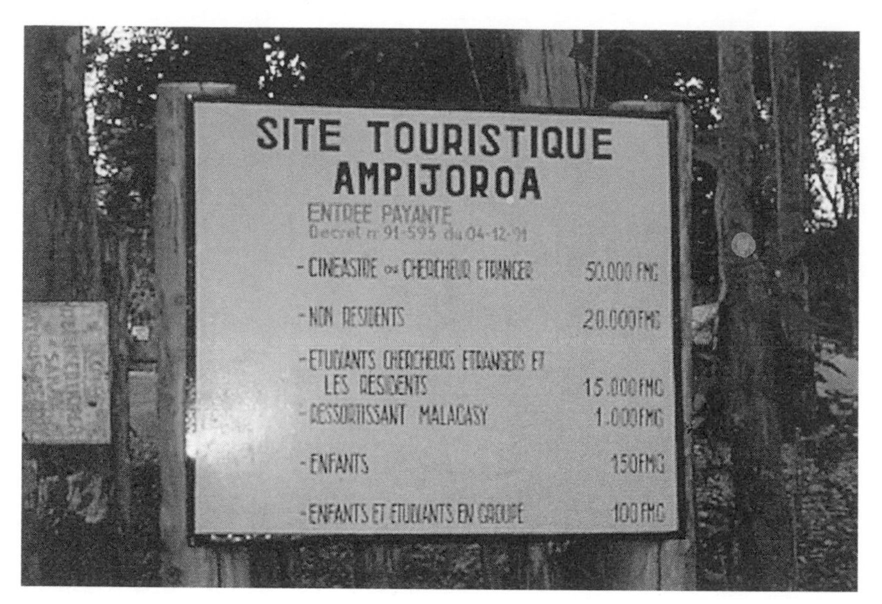

Fig. 2-3 List of admission fees at the entrance of Ampijoroa Forest Station. The entrance fee for non-nationals is about $10.

of the local staff who are knowledgeable about the flora and fauna, appear to be the most highly rated.

Fruit trees, such as mango, papaya, banana, jackfruit, and lime, are cultivated in the vicinity of the Ampijoroa village. Along Route 4 and the path stretching from the tourists' camping site to Lake Ravelobe, north of the camping area, tall, thick-trunked trees, such as the eucalyptus *Eucalyptus citriodora* and the thorny-trunked *Hura crepitans*, locally known as the crocodile tree (hazomvay in Malagasy), are conspicuous. During the transitional period between the dry and the rainy seasons-the seasons are described in the following section-large branches of these trees occasionally break and fall off. This is possibly because the moisture content of such trees increases drastically during this season.

The field study site of the Rufous Vanga

The field study site of our research on the Rufous Vanga is a 450 m × 550 m rectangular area called Jardin Botanique A (hereafter abbreviated to Jardin A) and the surrounding area. This area is located west of the forest station and is particularly suitable for ecological and behavioral studies because the trails are laid out in a grid pattern. Thus, researchers carrying a map are able to accurately

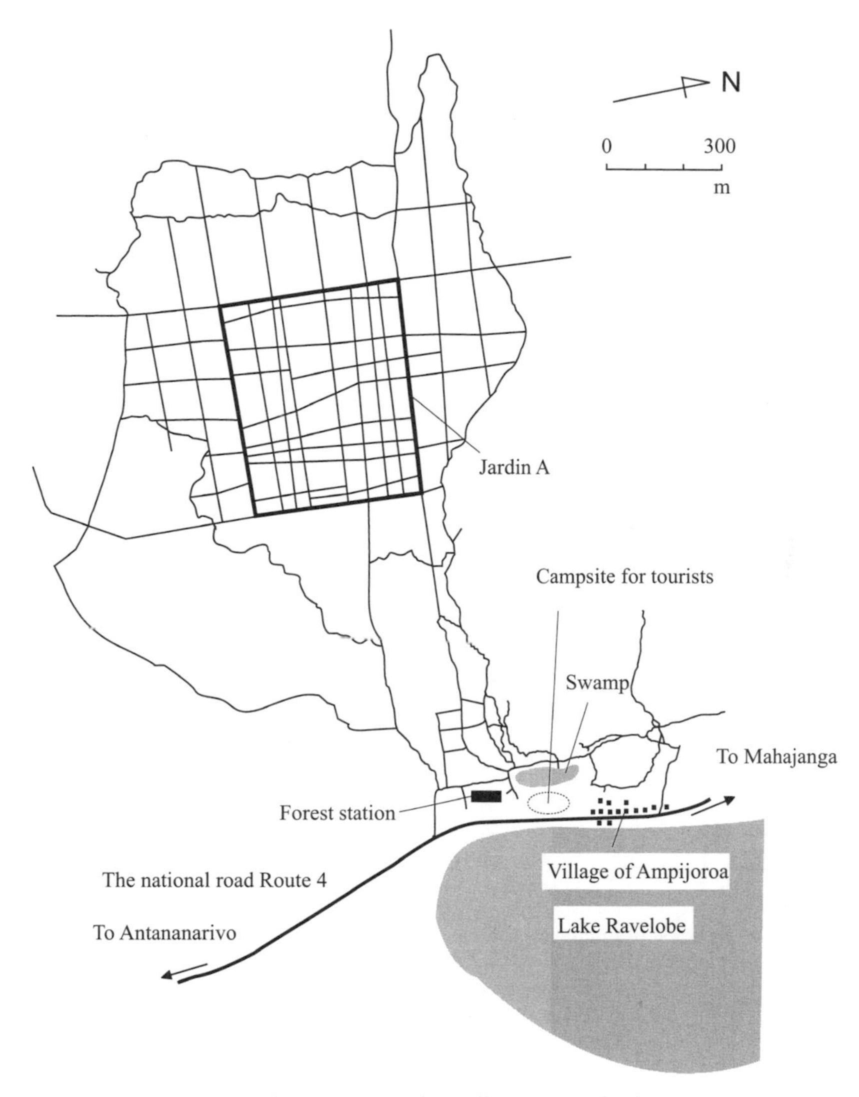

Fig. 2-4 Map of Ampijoroa Forest Station.

pinpoint their location and that of the organism they observe in the forest (Fig 2-4, Plate I-4).

Western Madagascar has two distinct seasons in a year-the rainy season and the dry season. Weather data collected systematically at Ampijoroa since 1996 indicates that the rainy season commences between late October and early December although the onset varies with the year (Fig 2-5). During the dry season, the daytime temperature exceeds 35° C but drops to around 15° C before

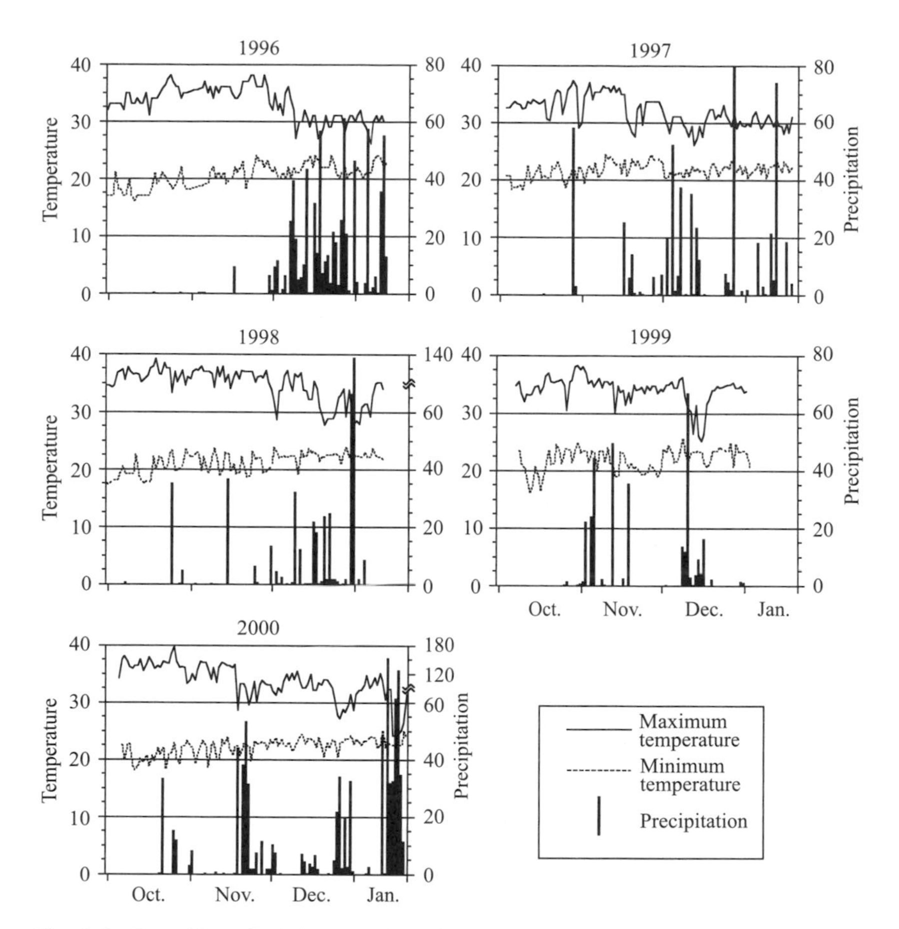

Fig. 2-5 Transition of minimum and maximum temperature, and that of precipitation, in Ampijoroa during the dry and rainy seasons from 1996 to 2000. Data in 1997, 1998, and 2000 is Mizuta's personal record. Data in 1996 is from U. Thalmann and A. M ller. Data in 1999 is by S. Asai.

morning dawns; thus, the diurnal temperature range is large. At the onset of the rainy season, the cloud cover increases, blocking sunlight. During this season, therefore, the maximum temperature does not show as high a rise as that during the dry season, and there is merely a marginal rise in the minimum temperature. During the dry season, humidity is low, approximately 20-60%, whereas it increases to approximately 60-90% during the rainy season.

The Ankarafantsika National Park is a forest located on a karst highland of limestone and characterized by a succession of gently undulating land and flat plains. On the whole, the area can be classified as a deciduous dry forest on the

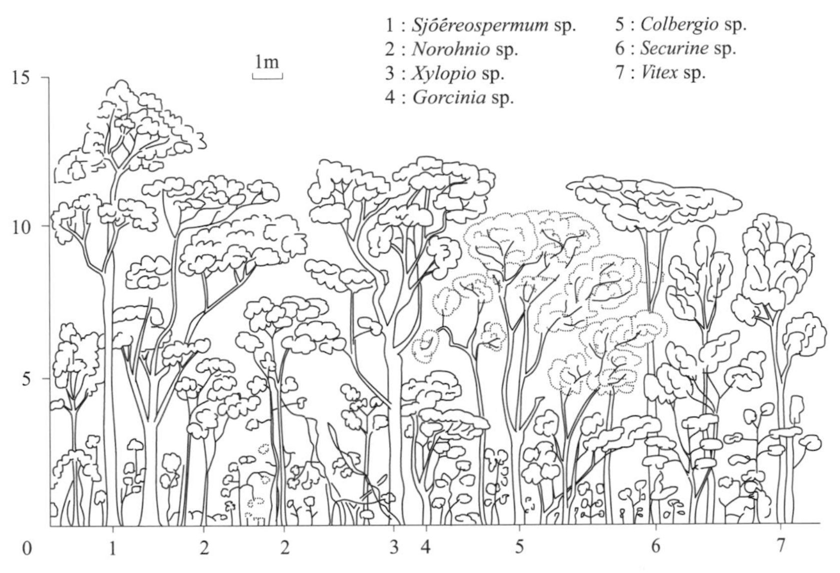

Fig. 2-6 Cross section diagram of the forest configuration in Ampijoroa (adapted from Razafy 1987[1]).

Table 2.1 Number of trees in the forest of Jardin A, according to the size of diameter at breast height (Hino 2002[2]).

Diameter (cm)	Number of trees/ha
1 – 5	23294.1
5 – 15	3455.5
15 – 30	336.1
30 –	53.8
	27139.5

basis of the vegetation found. The field study site in Ampijoroa, centered upon Jardin A, is also vegetated by a deciduous dry forest with a moderate canopy height of 10-15 m (Fig 2-6). The average height of the trees is approximately 12 m, and there are several thin trees of trunk diameter less than 5 cm at breast height (Table 2-1, Plates I-5 and II-1). Data on the vertical distribution of the vegetative cover in the forest indicates that undergrowth on the forest floor is relatively sparse, the canopy too is relatively less dense, and the maximum vegetation density is observed at the 5m level of the understorey (Fig 2-7). During the dry season, from April to around October, several trees shed their leaves, and the atmosphere of the forest is characterized by a greater degree of openness. On the other hand, the onset of the rainy season heralds the growth of new shoots and the forest becomes more verdant as the vegetation cover

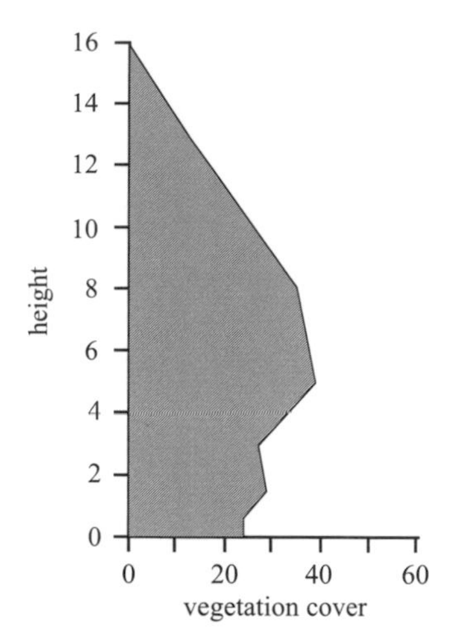

Fig. 2-7 Vertical distribution of vegetation cover in the forest of Jardin A (Hino 2002[2]).
Maximum tree height is 16m, and most trees are under 12m. Slender trees of which diameter at breast height is under 15cm are abundant, and the forest structure is comparable to the primary secondary-forest in Japan. Substratum vegetation is partly dense, however, on the whole it is scattered, therefore it is easy to walk around the area. Vegetation cover is concentrated at the height of 4m to 10m, which is favorable for tracking the behavior of birds which use canopy.

thickens daily (Plates I-2 and I-5). By December, the growth of tree foliage results in a marked decrease in visibility within the forest. Within the National Park, the existence of 56 families and 151 species of plants has been recorded. Woody plants account for a majority of these while creepers are relatively few in number.

The substrate of Jardin A and its vicinity consists of whitish sand, and the plants found there are not deep-rooted. Consequently, when heavy rain is accompanied by strong winds, even large trees are sometimes uprooted.

The forest vegetation on the fringes of the research facilities and Ampijoroa village is similar to that of Jardin A and its vicinity. However, perhaps as a result of previous forest fires, the trees found there are not as tall as those at Jardin A, and there is an abundance of bushes and low trees. The substrate of this area is a reddish soil, significantly different from that of Jardin A and its vicinity.

The birds of Ampijoroa

During the course of our research, which was conducted continuously from 1994-2000, 89 species of birds were recorded at Ampijoroa (see Appendix). Among these, 29 species were observed at Lake Ravelobe and its immediate vicinity, whereas the other species were observed primarily in the forests of Jardin A.

At the family level, the highest species diversity of birds in Ampijoroa is observed in the family Ardeidae (herons and egrets), which includes 11 species. The raptors, here comprised of the families Accipitridae and Falconidae, also included 11 species. All the herons and egrets were observed at Lake Ravelobe, whereas the raptors were found over a wide and diverse area, such as in the forest, around the lake, and elsewhere.

The Rufous Vanga, the main subject of this book, is a member of the family Vangidae. In Ampijoroa, the presence of seven species, half the total number of species in this family, has been recorded. Among all the families of Passeriformes (perching birds) observed in Ampijoroa, the family Vangidae is the most species rich. The striking ecological and morphological diversification in the family Vangidae, which is a splendid example of avian adaptive radiation[3], can be observed in this deciduous dry forest of Ampijoroa (Chapter 8).

Notably, among these, Van Dam's Vanga *Xenopirostris damii* occurs only in the Ankarafantsika National Park (Plate I-10). Van Dam's Vanga is considered to be one of the rarest and most endangered species within the family Vangidae because of its extremely specialized and limited habitat[4]. The ecology and behavior of this species were completely unknown until recent research conducted by our team documented its breeding biology[5].

Among the five species of the family Cuculidae, three species of Couas belonging to the same genus (Coquerel's Coua *Coua coquereli*, the Red-capped Coua *C. ruficeps*, and the Crested Coua *C. cristata*) occur sympatrically (i. e., they live in the same habitat). From the viewpoint of coexistence and adaptive radiation in birds, this species group is of particular interest. Comparative studies on these three species have revealed subtle differences in their respective foraging behaviors and their patterns of habitat use[6].

The White-breasted Mesite *Mesitornis variegata* and Schlegel's Asity *Philepitta schlegeli* are the other rare species found in the Ankarafantsika National Park that have been recorded at few other localities in Madagascar[4].

As noted above, Ankarafantsika is inhabited by numerous bird species, each occupying its own ecological niche. In comparison with the other regions of Madagascar, this region is noteworthy in terms of the number of extremely rare species abounding here. Ampijoroa is an invaluable forest, a biological treasure house, where birds including those of such rare species can be observed with relative ease. *Ny Vorona Malagasy*[7], a field guide to the birds of Madagascar written in Japanese and Malagasy, allows readers to glance at these birds in the comfort of their homes.

The breeding period of birds

A majority of the birds observed in Ampijoroa breed in the same area. In this context, the term "breeding" refers to a series of behaviors directed toward reproduction, and it includes activities such as courtship and mating, nest construction, egg laying, brooding, and chick rearing. Most birds have a fixed annual breeding period controlled by endogenous (internal) factors, such as hormone levels. However, exogenous (external) factors such as the quantity and quality of available food for maintaining the condition and activities of parent birds and chicks, the number of predators, and the seasonal changes in temperature and rainfall, also affect the timing of the breeding period of each species.

What exogenous factors affect the breeding period of the birds found in Ampijoroa? Figure 2-8 shows the breeding periods of birds in Ampijoroa, including those of several species of the family Vangidae. Only a portion of the total number of bird species breeding in this area is included; yet, it is evident that a majority of these species breed around and between November and December. These months herald the onset of the rainy season, at which time the numbers and activities of insects and small vertebrates increase dramatically; thus, for several bird species, the abundance of food is at its peak during this time. Since successful breeding requires tremendous energy, it is surmised that a large number of birds breed during this time because of the abundance of food.

Birds of the family Vangidae begin breeding from the end of September to mid-October, which is the end of the dry season. Their breeding period ends in early- to mid-January, the midst of the rainy season when it rains almost daily (Fig 2-8). During the rainy season, organisms that are potential food items for these birds increase in number; however, rainfall being a daily phenomenon, food capture is more difficult, and it takes longer to keep the chicks warm. Thus, the costs entailed in foraging and brood care increase. These increased costs may be reflected in the decrease in the number of breeding individuals observed during the rainy season. The reason why the Rufous Vanga completes its breeding during this time will be discussed in detail in Chapter 5. To ensure that the research spanned the entire breeding period of this species, the ecological and behavioral aspects of the Rufous Vanga, as described in this book, were

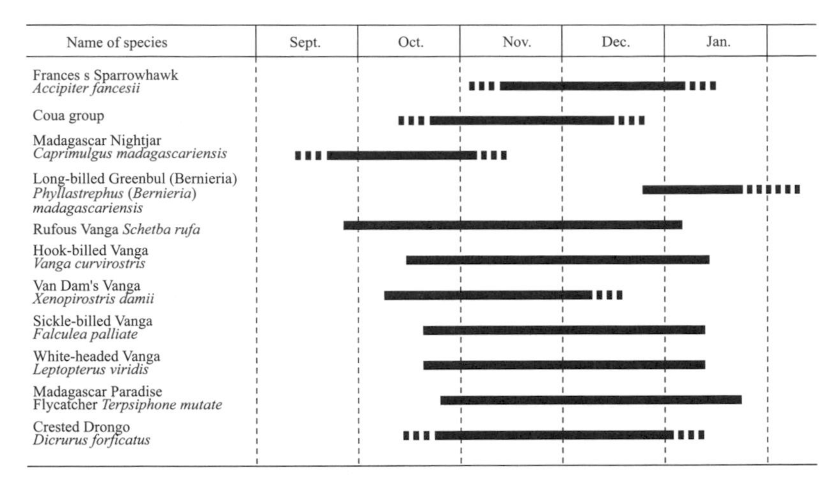

Fig. 2-8 Breeding seasons of main bird species observed in Ampijoroa.
Broken lines indicate the cases where the timing of the onset and completion of breeding is uncertain.

studied from late September until early January every year.

The breeding period of the Madagascar Paradise Flycatcher *Terpsiphone mutata*, the target species of my research, begins toward the end of October, slightly later than that of the family Vangidae (Fig 2-8). The Madagascar Paradise Flycatcher feeds primarily on flying insects, and these insects emerge in greater numbers during periods of frequent rainfall. This is probably the reason why this species begins breeding after the end of October when rainfall is certain[8]. Even within the Ampijoroa region, around the station and Lake Ravelobe, the breeding period of the Madagascar Paradise Flycatcher begins earlier than it does in the Jardin A region. The former region being nearer to water than the latter, several flying insects emerge here at an earlier period. The difference in the time and quantity of insect arrival between these two regions is believed to be the reason for the regional differences in the timing of the breeding period.

The diet of Frances's Sparrowhawk *Accipiter francesii* includes chicks of other birds. Although the breeding period of this bird was not specifically examined, it is known that they brood their eggs and raise their chicks in December (Fig 2-8). It has been suggested that this species times its breeding period to coincide with the period of maximum availability of the chicks of other birds with which it can feed its young.

Among the birds shown in Figure 2-8, I would particularly like to focus on the breeding period of two species-the Madagascar Nightjar *Caprimulgus madagascariensis* and the Long-billed Greenbul (Bernieria) *Phyllastrephus (Bernieria) madagascariensis*. The Madagascar Nightjar breeds earlier than other birds, beginning in September, whereas the Long-billed Greenbul begins breeding in late December when other birds have completed breeding. Why does the onset of the breeding period of these two species differ so markedly from that of other birds?

The Madagascar Nightjar does not construct a conspicuous nest; instead, it lays its eggs on the ground. The plumage of the adult is of a drab color similar to the dead leaves on the forest floor. Although this coloration serves to camouflage the adult, the eggs on the forest floor appear to be exposed and vulnerable to predators. Several snakes, such as the Colubrid *Leioheterodon madagascariensis*, crawl along the forest floor in search of food, and they will eat the eggs or chicks if they locate them. Parent birds that are incubating or rearing the chicks will also be in danger of being eaten. However, most of the snake species, including *L. madagascariensis*, emerge around November when rainfall becomes frequent. The Madagascar Nightjar avoids this risk of predation by completing breeding before the snakes emerge. Perhaps this is why this species begins breeding earlier than others.

Little is known about the Long-billed Greenbul. This species is often observed in Ampijoroa, feeding in groups. However, despite its abundance, we were unable to locate these birds' nests. Throughout the lower strata of the forest, we found several old, abandoned nests of unknown owner species. We supposed that these might have been the nests of the Long-billed Greenbul. However, we were never able to observe these nests while they were in use. Then, toward the end of December 2000, a few children in Ampijoroa village informed us that they had found a bird's nest, and they took us to see it. It was then that we saw, for the first time, a Long-billed Greenbul incubating in a nest. This nest was of the same type as the abandoned ones we had often found in the forest. We thereby confirmed that the breeding period of this species begins around this time, i. e., when the breeding period of other species is almost over. Why does this species breed so late in the season? Observation of the foraging behavior of the Long-billed Greenbul revealed that these birds hover around tree trunks and bushes in the lower and middle strata of the forest and use their long beaks to pick any insects that they discover[9]. This distinctive feeding method may be particularly effective during the rainy season when insects take shelter from the rain by concealing themselves behind barks and on the undersurfaces of leaves. If we assume that the foraging efficiency of this species increases when it rains, we may be able to explain why its breeding period begins around late December during the middle of the rainy season.

Clearly, the breeding of birds is closely associated with the life histories of those organisms that are part of their diet, the presence or absence of predators, and environmental factors such as temperature and precipitation levels. Therefore, to study the breeding ecology of birds, it is important to consider not only the study target species but also its relationships and interactions with other species.

Crocodile-infested Lake Ravelobe

Lake Ravelobe is located north of the Ampijoroa forest station on the opposite side of Route 4, which cuts through the forest station. The lake covers approximately 30 ha and is the habitat of numerous birds and other organisms.

As listed in the Appendix, several species of the heron and egret occur here; they forage in the shallow water along the lake's shoreline. The Madagascar Fish-Eagle *Haliaeetus vociferoides* also breeds in the forest along the lake shore, and occasionally, it can be seen perched on branches, looking down on the lake, or flying with its two-meter wingspan spread wide (Fig 2-9). The Madagascar Fish-Eagle is the largest raptor in Madagascar and is considered to be one of the rarest raptors in the world.

A tall eucalyptus tree alongside the lake is the communal roost site of the Sickle-billed Vanga *Falculea palliata*. A communal roost is a specific place that is utilized as a nighttime roosting site by several individuals of the same species. Within the family Vangidae, three species-the Sickle-billed Vanga, the White-headed Vanga *Leptopterus viridis*, and Chabert's Vanga *L. chabert*-display communal roosting behavior. The Sickle-billed Vanga is believed to use the same tree as a communal roosting site throughout the year, as does Chabert's Vanga. From mid-October to early December, the number of Sickle-billed Vangas using the communal roost decreases because breeding pairs leave the roost to build their own nests within breeding territories. Accordingly, it is surmised that the birds that stay behind at the communal roost during this period are unpaired non-breeding individuals. These non-breeding birds forage together in a group over a large area. From mid-December onward, the number of individuals in the roost greatly increases because of the influx of breeding individuals and young birds that were fledged during that year[10].

Lake Ravelobe is also inhabited by the Nile crocodile *Crocodylus niloticus*. During the daytime, it is often possible to observe several crocodiles basking on the beach or on the grassland near the water's edge (Plate I-6). When the shower rooms at the station were inaccessible due to construction work, we often bathed in this lake. It was refreshing to let the dust and sweat wash away in the lake, as we lay on our backs, looking up at the beautiful evening sky. Yet, it was also

Fig. 2-9 Madagascar Fish-Eagle *Haliaeetus vocifeeroides* which inhabits the lakeshore of L.Ravelobe.

rather disconcerting to see a crocodile cruising leisurely offshore. The local people advised us to disturb the surface of the lake before bathing in order to frighten off any crocodiles lying nearby. Despite this, a crocodile attack occurs at Lake Ravelobe at least once every few years. Nanga, a boy from the Ampijoroa village, who assisted us in locating birds' nests during the course of our research, was bitten by a crocodile and bears the scars of the attack on his back. Even more unfortunate was a tragic incident that occurred in October 2000, in which one of the station tour guides was killed by a crocodile in Lake Ravelobe. We were shocked and deeply affected when we heard of this incident because, as mentioned above, we often bathed in the lake. The locals told us that the Nile crocodiles in Lake Ravelobe are becoming increasingly aggressive, but I wonder at the extent to which this is true.

In 1994, when I first visited Ampijoroa, fishing in Lake Ravelobe with a net was forbidden and was almost considered a taboo. During those days, several people could be seen standing waist-deep in the water, using a fishing rod and line to fish (Plate I-7). However, recently, the number of people who fish in the lake with nets is said to have increased. I suppose it is possible that the increase in the catch of fish via netting might have resulted in a shortage of food for the crocodiles, forcing them to prey on humans approaching the water as an alternative. This could explain why the locals find that the Nile crocodiles have become more aggressive. In Madagascar, there exist numerous age-old taboos. Some of these taboos were probably based on indigenous knowledge and practices put in place to ensure that human activities would not disrupt the balance of the ecosystem. For humans to live harmoniously in the ecosystem, their activities should leave the ecosystem undisturbed. The people of Malagasy probably ensure that such activities are practiced through the imposition of taboos. Such a perspective may provide us with a compelling reason to address the present need to conserve nature.

Local folklore has it that in addition to crocodiles, mermaids live in Lake Ravelobe. This tale was recounted to me by Julien, a researcher at the Botanical and Zoological Park of Tsimbazaza and a collaborator in our project. At first, I thought I had misheard him and that he had probably mentioned the name of an unfamiliar animal. However, he had definitely said "mermaid," a mythical creature-half-human, half-fish. Lake Ravelobe is a freshwater lake; therefore, it is not possible that the marine-living dugong was being mistaken for a mermaid. (Reports of mermaid sightings by ancient sailors are believed to have originated from mistakenly identifying this marine mammal as a mermaid). Julien really meant a mermaid! Julien is an earnest ornithologist, however, he sometimes narrates truly incredulous stories with an absolutely straight face. In this instance, he recounted that one evening, as he was walking along the lake shore, he saw a mermaid sitting on a tree stump on the shore, washing herself. The mermaid soon realized that she was being watched, and she swiftly dived into the water. Julien claimed that a town of mermaids with numerous inhabitants existed beneath the waters of Lake Ravelobe. He also claimed that since it was most unusual for a mermaid to leave the water, he was extremely lucky to have witnessed this spectacle...

But I digress. There are several other ecosystems apart from the forest habitat. In comparison, humans and livestock are more constantly and actively interacting with these ecosystems than with the forest. This is particularly true of the lake. People living in the vicinity of the lake depend upon it for all the necessities of life, such as water for drinking and washing. Two piers on the lake shore are used for collecting water and washing clothes. In addition, paddy fields laid along the course of the stream that gushes from the lake are irrigated by the stream water. Thus, the lake water is used for agricultural purposes as well. Zebu cows, a valuable commodity in Madagascar, are maintained in pastures near the water (Fig 2-10). Particularly at the onset of the rainy season when the pastures turn green, the zebu cows graze heavily on the fresh grass, and the hump of fat on their back expands extensively. At times, the zebu cows are covered in mud, probably after having wallowed at the lake shore. Additionally, the fish found in the lake constitute an important source of protein for the locals. Lake Ravelobe

Fig. 2-10 Zebu cattle grazing on the lake shore of L. Ravelobe.

is thus an important resource not only for wild animals but also for the humans who reside in the vicinity. Apart from the image of crocodiles swimming by in the lake, the overall scenery of Lake Ravelobe, with people working in the paddy fields, cows grazing leisurely near the water, and birds walking in the shallow shoals hunting for food, is reminiscent of the "satoyama" (secondarily managed forest) landscape of Japan.

The life of researchers

Lastly, I will briefly describe a portion of the daily activities of the researchers who performed this study in the Ampijoroa forest.

As mentioned earlier, well-established accommodation facilities are available in the research field. During the research project, we stayed at this facility whenever possible. Unfortunately, the facility was often under construction or renovation. Moreover, the pace of construction was painfully slow. On several occasions, construction continued throughout the duration of that year's research period. At such times, when we were unable to use the accommodation provided by the facility, we lived in tents. Living under canvas for four consecutive months might seem challenging, but we used the tents only for sleeping in. Hence, it was not particularly uncomfortable or inconvenient. We spent most of the day outdoors. When not out on field research, such as during rest periods or when processing data, we used tables set under the roof of the accommodation or under a tarpaulin roof. Except for the time spent sleeping, we were in the open air all day and led an existence extremely unlike our indoor lifestyle in Japan.

Just how each researcher spent his day depended upon the research species that was the subject of the study. Their activities were dictated by the hours of activity they were required to spend upon their research subject. Most birds become active at dawn, and accordingly, our research began during the early hours of the morning. We awoke before 5 a. m., when it was still fairly dark. We did not undertake field research during the hot afternoons since birds grow less active as the temperature rises during the day. Instead, we spent these afternoon hours on other tasks. Toward late afternoon and evening, we resumed our observations, and we completed our studies before nightfall. Since we arose early every morning, we tended to sleep early as well, thereby leading a healthy lifestyle-early to bed, early to rise.

Researchers employed a few villagers from Ampijoroa as cooks and requested them to provide three meals each day. We requested Madame Jean, whose husband, Jackie, is regarded as one of the most experienced tour guides in Ampijoroa, to cook for us. Mme. Jean is a hard worker; at times, she cooked meals for as many as ten team members at one sitting without a word of complaint. It may come as a surprise to Japanese people that the staple food in Madagascar is rice. People in Madagascar eat an enormous quantity of rice with each meal. It would not be an exaggeration to say that with each meal, an adult consumes the equivalent of a washing-bowl full of rice. Hence, it is but natural that the per capita rice consumption in Madagascar is the highest in the world. Of course, the rice grown in this country is different from that grown in Japan. However, since it is not as dry as the long-grained rice of S. E. Asia, it is more palatable to the Japanese. A standard meal prepared by Mme. Jean consisted of a rice staple and side dishes of meat or fish and vegetable salad, followed by fruit. A soup made of beans and beef; ravintoto, which is a stewed dish prepared from cooked pork and shredded cassava leaves; fried chicken; and deep fried tilapia fish were some of the side dishes. Every one of these dishes was delicious, but the menu was rather limited. After a week, when the menu cycle began again, long-term visitors like us began to tire of the lack of variety. A small, hot chilli called sakay placed on the dining table served as a condiment. The mealtime drink called ranon'ampango-a kind of tea made from scorched rice boiled in water-is a tasty, crispy-flavored drink unique to Madagascar. Fruit is a cheap and important grocery item. The varieties of bananas and mangoes available differ according to the season and place of purchase, and it was interesting to compare the tastes of these different varieties. Fruit is an especially valuable food item because the other food tends to lack vitamins.

When research activity was not at its peak, we allowed ourselves a weekly holiday. On such days, we sometimes walked 4 km southeast along the national road to the Andranofasika village and shopped there. Every Wednesday, a market is held in the village and since early morning, the place teems with villagers. It was a pleasure to walk around the market, even if we did not make any purchases. The stands included those of a butcher and a fishmonger around which numerous flies gathered, brightly colored displays of vegetables and fruits, convenience-goods stalls, a garment shop selling suspiciously labeled designer clothing, a bicycle-parts shop, a tavoahangy (bottle) shop that sold empty bottles and cans, a lamba (cloth) shop, a chemist selling herbal medicines, and several other stalls. The variety of goods on display in the market ranged from items essential for daily life to those that we doubted anyone would ever find a use for. Upon tiring of the heat, we would visit a hotely (small restaurant) in the village, sip on a cold drink, and return to Ampijoroa by a minibus called "Taxi-brousse."

When we needed items unavailable at the Andranofasika market or when we required to exchange money at a bank, we had to travel two hours by car to the seaside town (by Malagasy standards) of Mahajanga. Lying next to the sea,

Fig. 2-11 A Mighty African baobab *Adansonia grandidieri* which grows at the intersection of Mahajanga.

Mahajanga has an open ambience (Fig 2-11). A company called Somapeche, which is an overseas affiliated company of the Maruha Corporation is located in this town. The Japanese staff of this company graciously conveyed various goods and letters to us. Mahajanga also hosts a good Chinese restaurant, a pizza shop, and an ice cream shop. A visit to these places was an incentive that helped us endure the two-hour car drive.

As mentioned earlier, Ampijoroa is an international meeting place where researchers from various countries convene. While research topics differed, we conducted research around the same general location; hence, we often exchanged information and discussed the position of our respective research projects. Whenever a team was about to leave for their home country, a farewell party was organized (Fig 2-12). Mme. Jean and other villagers prepared food for the party. We purchased a large quantity of beer and soft drinks and invited the staff of ANGAP as well as the villagers. Thus, each party was a fairly major event. The meals were always followed by dancing. Adults, children, and even

Fig. 2-12 Birthday and farewell parties are held for members of the international village for researchers.

elderly ladies danced in unique Malagasy styles to the loud local dance music blasting from a cassette recorder. The people of Madagascar are talented dancers and singers. I was impressed to witness even little children performing complicated steps in rhythm. Overseas researchers often attempted to learn Malagasy dance steps, and the dance party usually continued well into the night.

I have described only the amusements and the pleasurable activities that were part of our stay at the station. This is, of course, not reflective of our everyday life. In fact, for much of our time in Ampijoroa, we repeated the same research routine almost daily, thus leading a fairly monotonous existence with no entertainment. In reality, we sometimes suffered from stomach upsets because the water was of a different quality. Moreover, we were often bothered by mosquitoes and horseflies as well as tiny sweat bees that hovered in dozens around our face to lap up sweat. The daytime heat was stifling, and the rain that occasionally persisted for several days often led us to curse the weather. It was against this background that we lived our daily life, and depending on the progress of our research, we fluctuated between enthusiasm and despair. Even within our team, several members contracted malaria (Fig 2-13). We have sometimes been the target of envy for having conducted our research in Madagascar; however, in reality, the research itself was a boring activity which demands much perseverance.

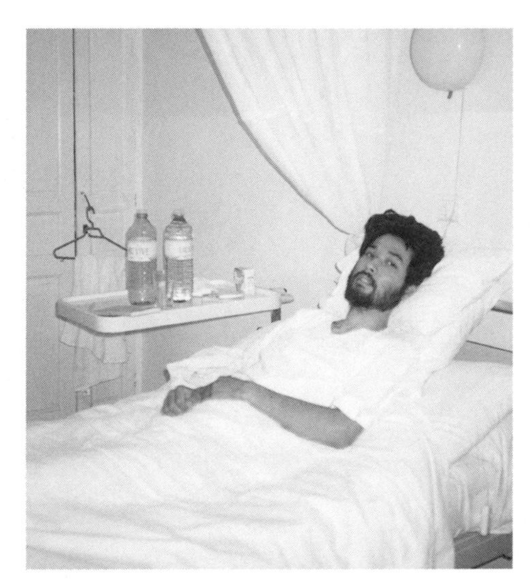

Fig. 2-13 Dr. S. Asai, confined to hospital at Antananarivo, capital of Madagascar, suffering from a severe bout of malaria. Fine medical service is provided in the hospital.

In the 10 years since we began studying there, Ampijoroa has undergone several changes. In 1990, when Dr Yamagishi and Nakamura, then a graduate student, visited Ampijoroa as the pilot Japanese team for this research project, there were no facilities available and they lived in a tent, even plucking and quartering chickens on their own. As the research facilities improved, the number of overseas researchers seeking to visit Ampijoroa has increased. Simultaneously, the development of a tourist system also resulted in an increase in the number of tourists. It is important that care be taken to ensure that the increased number of visitors does not unduly disturb animals and plants in the forest. I believe that the attitude of the villagers of Ampijoroa toward the forest has changed with the increase in the number of overseas visitors. I consider it necessary for the people concerned, namely the staff of ANGAP, villagers living near the forest, researchers, tourists, and others, to determine the best means of preserving this precious forest.

The name Ampijoroa is derived from the word "fijoroana," which means "a place to conduct joro." (Joro is a religious ritual.) "Joro" is directed toward invoking the blessings of God, sacred ancestors, and living people[11]. Thus, Ampijoroa connotes a sacred area. Indeed, on the northern shore of Lake Ravelobe, there exists a place where joro is performed (Fig 2-14). It is probable that the locals designated the lake and its vicinity, including this sacred site, as a

Fig. 2-14 A prayer ritual takes place on the lakeshore of L. Ravelobe, entreating ancestors to bless them. Zebu cattle are driven into the corral, slaughtered, and consecrated to crocodiles.

holy area. Over 90% of the forests of Madagascar have disappeared because of human activities. Given this situation, the fact that Ampijoroa has retained an extensive forest area could be attributed to the reluctance of the local people to cut down the trees within their sacred site. I sincerely hope that the Ampijoroa forest will be considered a sacred area in the literal sense and that it will remain unspoiled, supporting not only humans but also numerous living organisms, including the Rufous Vanga.

References

1 Razafy, F. L. (1987) La Réserve Forestiére d'Ampijoroa: Son modéle et son bilan. Mémoire de fin d'études, Université de Madagascar.

2 Hino, T. (2002) Breeding bird community and mixed-species flocking in a deciduous broad-leaved forest in western Madagascar. *Ornithological Science*, 1: 111-116.

3 Yamagishi, S. & Eguchi, K. (1996) Comparative ecology of Madagascar vangids (Vangidae). *Ibis*, 138: 283-290.

4 Langrand, O. (1990) *Guide to the Birds of Madagascar*. Yale University Press, London.

5 Mizuta, T., Nakamura, M., & Yamagishi, S. (2001) Breeding ecology of Van Dam's Vanga *Xenopirostris damii*, an endemic species in Madagascar. *Journal of the Yamashina Institute for Ornithology*, 33: 15-24.

6 Urano, E., Yamagishi, S., Andrianarimisa, A., & Andriatsarafara, S. (1994) Different habitat use among three sympatric species of couas *Coua cristata, C. coquereli* and *C. ruficeps* in western Madagascar. *Ibis,* 136: 485-487.
7 Yamagishi, S., Masuda, T., & Rakotomanana, H. (1997) *Ny Vorona Malagasy*. Kaiyusha, Tokyo (in Japanese and Malagasy).
8 Mizuta, T. (2002) Seasonal changes in egg mass and timing of laying in the Madagascar Paradise Flycatcher *Terpsiphone mutata. Ostrich*, 73: 5-10.
9 Eguchi, K., Yamagishi, S., & Randrianasolo, V. (1993) The composition and foraging behaviour of mixed-species flocks of forest-living birds in Madagascar. *Ibis,* 135: 91-96.
10 Eguchi, K., Amano, H. E., & Yamagishi, S. (2001) Roosting, range use and foraging behaviour of the Sickle-billed Vanga *Falculea palliata* in Madagascar. *Ostrich*, 72: 127-133.
11 Moriyama, T. (1996) *Tombs and Social Practice among the Sihanaka of Madagacar, An Ethnographic Study*. University of Tokyo Press, Tokyo (in Japanese).

Annexed List — Birds observed in Ampijoroa during the study periods between 1994 and 2001 —

No.	Common name	Scientific name
	Family: Cormorants — Phalacrocoracidae	
1	Long-tailed cormorant	*Phalacrocorax africanus*(R)
	Family: Anhingas and Darters — Anhingidae	
2	Oriental darter	*Anhinaga melanogater*(R)
	Family: Herons — Ardeidae	
3	Grey heron	*Ardea cinerea*(R)
4	Madagascar heron and humblot's heron	*Ardea humbloti*(R)
5	Purple heron	*Ardea purpurea*(R)
6	Great egret	*Casmerodius albus*(R)
7	Black heron	*Egretta ardesiaca*(R)
8	Cattle egret	*Bubulcus ibis*(R)
9	Dimorphic egret	*Egretta dimorpha*(R)
10	Squacco heron	*Ardeola ralloides*(R)
11	Madagascar pond-heron	*Ardeola idae*(R)
12	Striated heron and little heron	*Butorides striatus*(R)
13	Black-crowned night-heron	*Nycticorax nycticorax*(R)
	Family: Ibises and Spoonbills— Threskiornithidae	
14	Glossy ibis	*Plegadis falcinellus*(R)
15	Madagascar crested ibis	*Lophotibis cristata*
	Family: Eagles, Hawks, and Kites — Accipitridae	
16	Bat hawk	*Machaeramphus alcinus*
17	Black kite	*Milvus migrans*(R)
18	Madagascar fish-eagle	*Haliaeetus vociferoides*
19	Madagascar farrier hawk	*Polyboroides radiatus*
20	Frances' sparrowhawk	*Accipiter francesii*
21	Madagascar sparrowhawk	*Accipiter madagascariensis*
22	Henst's goshawk	*Accipiter henstii*
23	Madagascar buzzard	*Buteo brachypterus*
24	Madagascar kestrel (Newton's kestrel)	*Falco newtoni*
25	Banded kestrel	*Falco zoniventris*
26	Sooty falcon	*Falco concolor*
	Family: Ducks, Ducks and Geese, Geese, and Swans — Anatidae	
27	White-faced whistling-duck	*Dendrocygna viduata*
28	Comb duck	*Sarkidiornis melanotos*(R)
	Family: Guineafowl — Numididae	
29	Helmeted guineafowl	*Numida meleagris*(R)
	Family: Mesites — Mesitornithidae	
30	White-breasted mesite	*Mesitornis variegata*
	Family: buttonquail — Turnicidae	
31	Madagascar buttonquail	*Turnix nigricollis*
	Family: Coots, Gallinules, and Rails — Rallidae	
32	White-throated rail	*Dryolimnas cuvieri*(R)
33	Allen's gallinule	*Porphyruda alleni*(R)
	Family: Jacanas — Jacanidae	
34	Madagascar jacana	*Actophilornis albinucha*(R)
	Family: Painted Snipe — Rostratulidae	
35	Greater painted snipe	*Rostratula benghalensis*(R)

Family: Avocets, Lapwings, Oystercatchers, Plovers, and Stilts — Charadriida	
36 Three-banded plover	*Charadrius tricollaris*(R)
Family: Sandpipers and Sanpipers — Scolopacidae	
37 Common greenshank	*Tringa nebularia*(R)
38 Common sandpiper	*Actitis hypoleuncos*(R)
Family: Sandgrouse — Pteroclididae	
39 Madagascar sandgrouse	*Pterocles personatus*(R)
Family: Doves and Pigeons — Columbidae	
40 Madagascar turtle-dove	*Streptopelia picturata*
41 Namaqua dove	*Oena capensis*
42 Madagascar green-pigeon	*Treron australis*
Family: Parrots — Psittacidae	
43 Vasa parrot	*Coracopsis vasa*
44 Black parrot	*Coracopsis nigra*
45 Grey-headed lovebird/Madagascar lovebird	*Agapornis cana*
Family: Cuckoos, Roadrunners, and Relatives — Cuculidae	
46 Madagascar cuckoo	*Cuculus rochii*
47 Coquerel's coua	*Coua coquereli*
48 Red-capped coua	*Coua ruficeps*
49 Crested coua	*Coua cristata*
50 Madagascar coucal	*Centropus toulou*
Family: Barn Owls, Barn-owls, Masked Owls, and Relatives — Tytonidae	
51 Barn Owl	*Tyto alba*
Family: Typical Owls — Strigidae	
52 Malagasy scops-owl	*Otus rutilus*
Family: Nightjars — Caprimulgidae	
53 Madagascar nightjar	*Caprimulgus madagascariensis*
Family: Swifts-Apodidae	
54 African palm-swift	*Cypsiurus parvus*
Family: Kingfishers — Alcedinidae	
55 Madagascar kngfisher	*Corythornis vintsioides*
56 Madagascar pygmy-kingfisher ???	*Ispidina madagascariensis*
Family: Devilfishes, Goblinfishes, and Stonefishes — Synancejidae	
57 Onestick stingfish	*Merops superciliosus*
Family: Rollers — Coraciidae	
58 Broad-billed roller	*Eurystomus glaucurus*
Family: Cuckoo-roller — Leptosomidae	
59 Cuckoo roller and courol	*Leptosomus discolor*
Family: Hoopoes — Upupidae	
60 Eurasian hoopoe	*Upupa epops*
Family: Asity and False Sunbird — Philepittidae	
61 Schlegel's asity	*Philepitta schlegeli*
Family: Swallows — Hirundinidae	
62 Mascarene martin	*Phedina borbonica*

Family: Cuckoo-shrikes — Campephagidae	
63 Ashy cuckooshrike	*Coracina cinerea*
Family: Bulbuls — Pycnonotidae	
64 Long-billed Greenbul (Bernieria)	*Phyllastrephus (Bernieria) madagascariensis*
65 Madagascar bulbul	*Hypsipetes madagascariensis*
Family: Vangas — Vangidae	
66 Rufous vanga	*Schetba rufa*
67 Hook-billed vanga	*Vanga curvirostris*
Family: Crows, Jays, and Magpies — Corvidae	
68 Van Dam's vanga	*Xenopirostris damii*
69 Sickle-billed vanga	*Falculea palliate*
70 White-headed vanga	*Leptopterus viridis*
71 Chabert's vanga	*Leptopterus chabert*
72 Blue vanga	*Cyanolanius madagascarinus*
Family: Thrushes — Turdidae	
73 Madagascar magpie-robin	*Copsychus albospecularis*
Family: Babblers — Timaliidae	
74 Common jery	*Neomixis tenella*
Family: Gnatcatchers, Kinglets, and Old World Warblers — Sylviidae	
75 Madagascar swamp-warbler	*Acrocephalus newtoni*(R)
76 Malagasy brush-warbler	*Nesillas typica*
77 Common newtonia	*Newtonia brunneicauda*
Family: Monarch Fycatchers — Monarchidae	
78 Madagascar paradise-flycatcher	*Terpsiphone mutata*
Family: Sunbirds — Nectariniidae	
79 Souimanga sunbird	*Nectarinis souimanga*
80 Long-billed green sunbird	*Nectarinia notata*
Family: White-eyes — Zosteropidae	
81 Madagascar white eye	*Zosterops maderaspatana*
Family: Estrildid Finches — Estrildidae	
82 Madagascar munia	*Lonchura nana*
Family: Bishops, Weaver Finches, and Weavers — Ploceidae	
83 Sakalava weaver	*Ploceus sakalava*
84 Madagascar red fody	*Foudia madagascariensis*
Family: Drongos — Dicruridae	
85 Crested drongo	*Dicrurus forficatus*
Family: Crows, Jays, and Magpies — Corvidae	
86 Pied crow	*Corvus albus*(R)

(R) Birds oberved around the Lake Ravelobe

Basic biology of the Rufous Vanga

Satoshi Yamagishi and Shigeki Asai

Observing the Rufous Vanga

Rufous Vangas are found in and around the Ampijoroa village and can be easily observed there. The call of the Rufous Vanga consists of three different notes, "Kwa, Kwa, Kwoo," with a falling intonation (Fig 3-1), and it penetrates deep into the forest. If you try to locate these birds by tracing the source of their call, you will notice them perched on thick branches of trees, just above eye-level. Since they seldom sit on thin branches of trees and are not so afraid of humans, they can be observed from close quarters. The Rufous Vanga is a small bird with a tri-colored plumage: black, on the head; white, on the stomach; and reddish brown, on the back. The reddish brown plumage that covers the back of the Rufous Vanga is not too bright. In fact, it more closely resembles the color of withered leaves (refer to the picture on the book jacket). Total length of this bird measures 20 cm in length[1]. Although images and illustrations of Rufous Vangas in books convey the impression that their plumage is brightly colored, these birds are surprisingly inconspicuous in the forest.

There exists a clear sexual differentiation in plumage coloration: in the male, the entire head and the area from below the throat to the chest are black (Plate I-1b, I-11b); whereas in the female, only the area from the head to just below the eyes is black, while the cheeks and the area from the throat to chest are white (Plate I-1c, I-11c).

After locating an individual Rufous Vanga and observing it, it is quite common to see a second bird appear in a while; this would be the first bird's breeding mate (Plate I-9). Thereafter, the observer can follow the pair of birds flying through the trees, each bird advancing ahead alternately—a male, which is black from the throat to the chest and a female, which is white in the

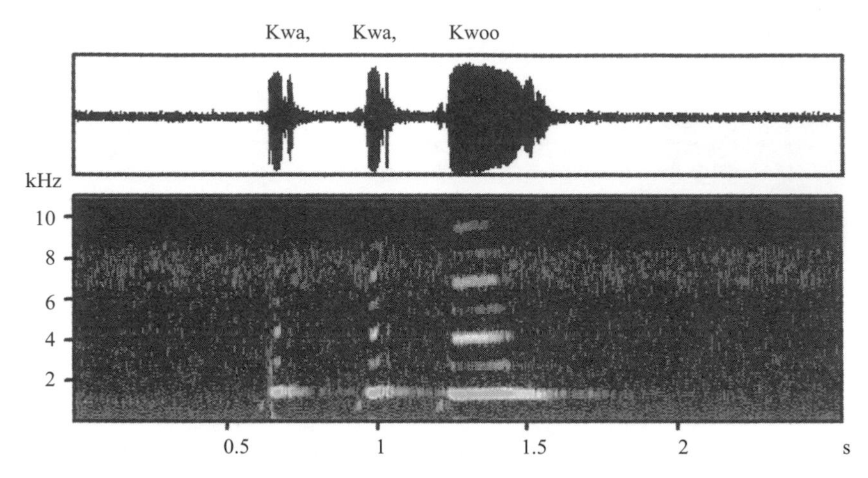

Fig. 3-1 Typical chirping notes, three syllables of, Kwa, Kwa, Kwoo, (Recording: Yukio Takeda, sonogram preparation: Eri Honda)

corresponding area. While foraging, these pairs do not cover a considerable distance in one flight. Thus, an observer who loses sight of them will usually be able to trace them upon moving approximately 10 meters in the direction in which they were flying when they disappeared from view. These birds land on the ground and forage for prey such as crickets, centipedes, and scorpions. Although most of their prey is captured on the ground[2], they also forage arboreally for caterpillars of butterflies and moths, and cicadas during the period in which these insects are abundant.

During our search for and observations of a breeding pair, we sometimes encountered a third and a fourth individual within the territory. These birds were helpers, as has been mentioned in Chapter 1. There are two types of helpers: one with a black throat and apparently the same plumage pattern as a paired male and another with a spotted throat (Plate I-1a, I-11a). During the six years of our research (1994-1999), we banded 294 chicks while they were in the nest to enable individual identification of each bird and then traced the subsequent changes in their plumage pattern. Among these birds, all 51 males that were identified the following year as yearlings (one-year olds) had spots on their throat, confirming that the spotted-throat birds were yearling males, as we expected[3]. The throats of yearling males are white, similar to those of females, but the presence of the black spots enabled us to distinguish them from females. On rare occasions, we also encountered some yearling males that had a blackish throat, thereby resembling the paired male because the spots on the chest extended into the black throat area (Plate I-8). By age two, the throats of such

Table 3-1 The number of male chicks which were banded, and the transition of plumage color of their throats (from Yamagishi et al. (2002)[3]).

Research Year	Total number of chicks which were banded	Number of males observed					
		1995	1996	1997	1998	1999	2000
1994	14	2	2	2	2	2	2
1995	38		8	7	7	6	3
1996	68			15	12	10	6
1997	53				10	7	7
1998	76					10	7
1999	45						6
Total	294						

Numbers with shading are birds observed as spotted-throat individuals. Others were black-throat individuals.

spotted-throat individuals become completely black; thereafter, they never revert to the spotted pattern (Table 3-1)[3]. Hereafter in this book, we will refer to the spotted-throat individuals as "yearling males," and to males with black throats as "males aged two or over," or "adult males." As opposed to the transformation observed in males, females retain the plumage coloration observed in their first year. We were able to reach these specific conclusions as a result of the leg-banding exercise we undertook, as described in the next section.

Leg banding birds for individual identification

In order to elucidate the ecological and behavioral features of the Rufous Vanga, it was necessary to identify the individual observed in the field. For instance, where did a helper that aids in breeding activities come from? To answer this question, we invested our best efforts in identifying each of the Rufous Vangas inhabiting our study area. Since it was impossible for us to distinguish among Rufous Vangas by any unique visual features, which is how we recognize fellow humans, we banded the legs of each bird with a unique combination of colored bands instead. This enabled us to identify the bird being observed by simply noting the color of the bands on its legs. However, to band an individual, it was first necessary to capture it.

We caught birds using mist nets—a type of net made from fine black thread. The net is positioned vertically across the flight path of birds. When birds fly into the net, they get entangled in the slack pockets of the net and are captured (Fig 3-2). Mist nets are commonly used as a tool to capture small birds; however, in the extended forest habitat where our studies were conducted, birds

Figure 3-2 Adult birds are trapped using a mist net.

are not so easily caught in mist nets. The location at which a mist net must be set is decided after confirmation of their main area of activity and their most frequently-used flight routes by careful observation of the targeted individuals.

In the case of the Rufous Vanga, when we played the recorded sound of their own chirps in their territory, paired birds and helpers flocked together and clamorously chattered around the sound emanating from the speaker in order to scare away the perceived intruders. Taking advantage of such behavior, we were often able to capture Rufous Vangas by replaying their chirping sound in the vicinity of a mist net. The Rufous Vanga is extremely aggressive by nature. Trapped individuals would emit piercing screeches whenever we tried to free them from the net. Group members sharing the same territory are closely united and do not fly away after they hear the screeching sound of the trapped individual. They approach and fly around the trapped bird, eventually getting caught in the net themselves, one after another.

After leg banding a bird for individual identification, measuring the size of each body part, and collecting a 0.05 ml or so blood sample from the vein on the underside of the wing (Fig 3-3), the captured bird was released. The bands used to enable recognition of individuals are made of plastic and consist of several colors. Each Rufous Vanga was banded with two bands on one leg. Combining an upper and lower band from 10 kinds of colors allow us to identify 100 individuals. Using differently combinations on the right and left legs,

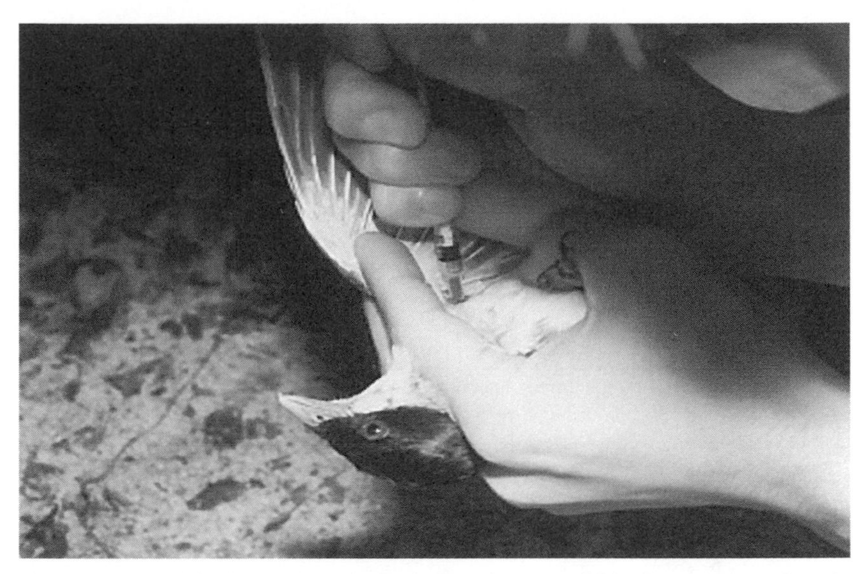

Fig. 3-3 Trapped birds are released after being taken a small blood sample from the vein beneath the wing, and being banded with color rings.

respectively, would have enabled us to uniquely identify 10,000 individuals. However, after experimentally applying this method to some individuals, we found it extremely difficult to recognize the color combinations of both legs simultaneously. We therefore used bands with a similarly paired color combination for both legs, facilitating identification of an individual by checking a single leg. Thereafter, we called the color combination of their bands as a name of each individual.

One of the most significant pieces of information that can be obtained by capturing a bird is its age. As also observed in certain other passerine birds, the color of their wings at age one is different from that at age two and over. One-year-old Rufous Vangas have a tinted plumage on the upper surface (of their backside) and tinted (buff color) portions on the tips of their wing coverts (Frontispiece I-14). Thus, even female Rufous Vangas, which do not have spotted-throat patterns, can be distinguished as yearlings upon close examination. Unfortunately, it has not yet been possible to accurately assess the age of individuals, that are over two years old when captured.

Table 3-2 presents the values obtained from measurements of captured birds. In the present context, the term "larger" denotes a statistically significant difference. The data indicates that the males are larger than females in all dimensions, except in terms of weight. However, no sexual difference was evident in the tarsus length of yearlings. Considered together, the data indicates

Table 3-2 Measurements of body parts of the Rufous Vanga ± standard deviation (the number of samples)

	Wing length (mm)**	Tail length (mm)**	Tarsus length (mm)*	Bill length (mm)**	Body weight (g)
Male of age two and over	106.8 ± 2.5 (134)	85.5 ± 3.2 (133)	24.5 ± 0.9 (135)	26.4 ± 2.1 (135)	41.7 ± 2.4 (134)
Female of age two and over	103.2 ± 6.2 (72)	83.5 ± 3.4 (68)	24.0 ± 0.9 (72)	25.5 ± 1.6 (70)	41.6 ± 3.4 (68)
One-year-old male	103.5 ± 2.4 (38)	85.1 ± 3.9 (37)	24.4 ± 0.6 (38)	26.9 ± 1.7 (38)	40.0 ± 2.2 (37)
One-year-old female	101.3 ± 2.1 (25)	82.5 ± 2.5 (25)	24.1 ± 0.7 (25)	25.6 ± 1.6 (25)	39.1 ± 3.0 (25)

Individuals which were caught more than once are included.

**: In both one-year-old and two-year-old and over individuals statistically significant sexual dimorphism was evident.

* : A statistically significant sexual dimorphism was evident only in two-year-old and over individuals.

In yearling and two-year-old and over males and females, a statistically significant difference was seen in wing length and body weight.

the presence of a sexual dimorphism in the Rufous Vanga, with males being larger than females. However, this difference is slight: even in terms of wing length, which shows the largest difference among all measurements taken, the sexual difference is only 3 mm (approximately 3%). In wing length and body weight, there exists a difference between birds that are yearlings and those that are two years old and above. However, no difference was evident in other body parts. Consequently, we can ascertain that the Rufous Vanga achieves adult size at age one.

Since the primary focus of this period of research was breeding ecology, our efforts were concentrated particularly on nest finding and observation, and we recorded the process of parenting at each nest every year. The future whereabouts of chicks born in a nest are critical data; therefore, we took approximately six-day-old hatchlings out of the nest, leg banded them with colored bands, measured their body parts, and then replaced them in the nest (Fig 3-4, Frontispiece I-15). Simultaneously, we collected a 0.05 ml or so blood sample from the chicks for sexing through DNA examination (Fig 3-5). This practice enabled the identification of most of the individuals while they were still chicks; thus, by the time they fledged, they had already been leg banded with rings, in other words, they had been name-tagged. Over the course of a seven-year period, approximately 500 individuals, including adults trapped with mist nets and chicks found in nests, were leg banded for identification. This leg-banding operation for the identification of individuals was the foundation of all our research activities.

Fig. 3-4 Dr. K. Eguchi brings chicks down from their nest, using a simple tree-climbing ladder.

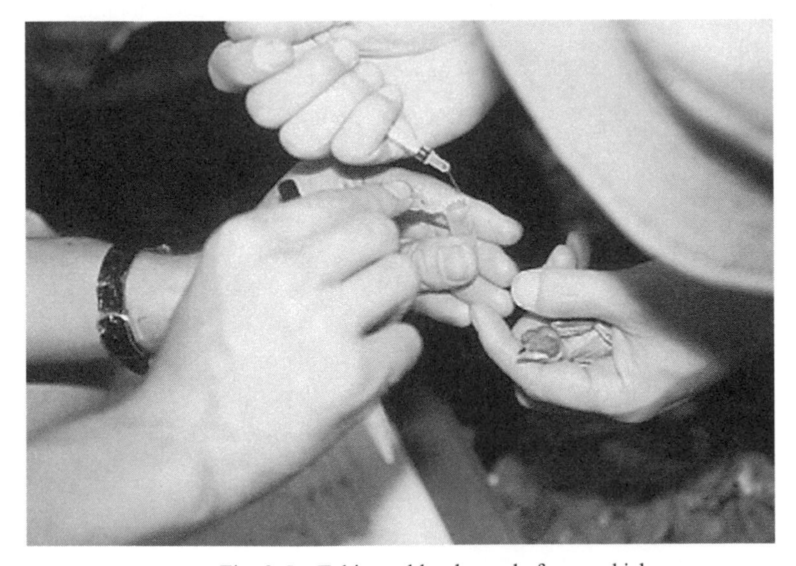

Fig. 3-5 Taking a blood sample from a chick.

The individual banded with pink stripes

The individual banded with pink stripes, as mentioned in Chapter 1, had a spotted-throat in 1991. As we predicted, it had developed into a male with a black throat in the following year, and it turned into a breeding male and settled in a territory adjacent to that in which it had been a helper. We learned of this in 1992 from the British scientist Dr Hawkins while we were in Japan, frustrated due to our inability to obtain funds for our project from the Japanese Ministry of Education, Culture, Sports, Science and Technology.

Eventually, we were awarded a research grant, and in 1994, we visited Madagascar again in high spirits. The individual banded with pink stripes was still occupying the territory that Dr Hawkins had described to us. At that time, the male was already four years old. We termed its territory "Group-E." It was inhabited by four individuals: the pink-stripe banded individual and its mate, and two helpers—one male and one female (Fig 3-6). Toward the end of the breeding season, the female helper moved to Group-H. During the non-breeding period in 1994-95, the male helper became independent and left to establish a new territory, Group-EE. Moreover, in 1994, the pink-stripe banded male and its mate failed to breed; therefore, during the next breeding season in 1995, only the two pair members remained in their territory (Fig 3-6). In 1996, they finally succeeded in fledging a single male, and this individual became a spotted-throat helper in 1997 when two males were fledged. In 1998, the spotted-throat male developed into a black-throated male and became independent, founded Group-HH, and formed a new pair with a female from a different group. In 1998, the two males produced in Group-E during the preceding year developed into spotted-throat males and helped their parents; however, no chicks were successfully fledged that year. In 1999, the two spotted-throat males turned into black-throated males and helped the parents. Once more, however, no chicks were fledged. Subsequently, the female parent moved to Group-FX, and in 2000, one of the black-throated males founded Group-IY and obtained a mate. Meanwhile, another female had recently moved to Group-E, which now consisted of a female and two males (Fig 3-6). In 1995, Group-TX was established from EE, which had been partitioned from E. The location of these partitioned territories is shown at the lower left portion of Figure 3-6, which portrays the life cycle of the pink-stripe banded individual until it reached 10 years of age. This figure encapsulates the history, development, and spread of the family.

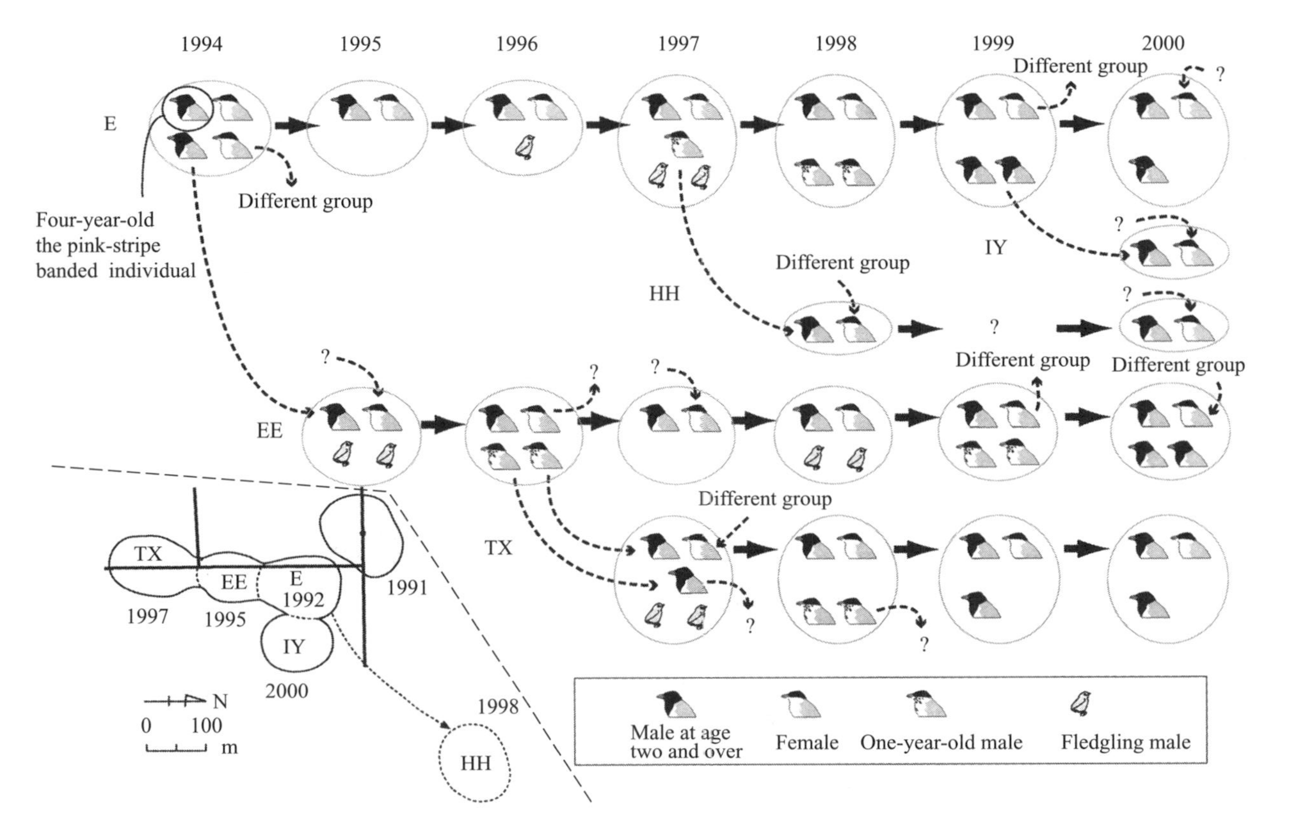

Fig. 3-6 The thriving family lineage of the pink-stripe banded male individual. For chicks, only males are shown. Refer to Fig. 3-7 in detail.

The number of birds in a group

It might appear that we are focusing excessively on the pink-stripe banded male, which is but one individual among many. However, by exemplifying the pink-stripe banded male, the basis of the social structure of the Rufous Vanga can be clearly illustrated. Figure 3-6 depicts several aspects of this society, such as daughters leaving their natal territory, the fact that most helpers are sons, that only the original pair remains after all the sons leave, and that when they have another son, the group becomes a single female/multiple male group the following year. These features are clearly depicted in Figure 3-6. An examination of the history of a group shows that the pattern of single male/single female pair and a group of single female/multiple males recurs. A closer look at this phenomenon reveals a social structure wherein both pairs and single female/multiple male groups are present in any family lineage at different periods. In what ratio do these types of groups occur? Further, what is the average number of individuals within a group of Rufous Vangas?

For those readers who do not like individual examples, Figure 3-7 (the folded figure) graphically illustrates the changes across all groups examined within the study area over a seven-year study period[4]. We hope that this figure will enable readers to appreciate the number and makeup of individuals that comprised a group, identify the individuals and the group to which they moved, and also identify when they moved. Firstly, from Figures 3-7 and 3-8, it is evident that females often move between groups, while males seldom move from a group in which they settle. However, as mentioned in Chapter 2, our research period was restricted to about half a year, from September to the following January. It was thus impossible for us to monitor events that occurred between the time we returned to Japan and our next visit, i. e., between the end of one breeding period and just before the onset of the following one. This was the weak link in our research. Most of the movement of individuals between groups occurred during our absence; however, there was no major difference in the number of those groups that had been mentioned before.

Table 3-3 indicates the number of individuals categorized into each status across 251 groups from 1994-2000. It is evident that 63% of Rufous Vanga groups consisted of pairs and that 37% of these groups consisted of single female/multiple males, namely, a pair and several male helpers. Sixteen percent of the groups contained only yearling spotted-throat individual helper(s) (average 1.3 individuals/group), whereas 5% of the groups also included two-year-old or over black-throated male helper(s) (average 1.4 yearling individuals/group and 1.3 two-year-old or over individuals/group). Except for

Table 3-3 Comparison of group composition
The lowest column indicates the number of helpers per group. Group composition at the end of the breeding season is shown.

Research year	Number of group	Breeding pair only	Breeding Pair +		
			One-year-old male	Two-year-old and over male	One-year-old male + two-year-old and over male
1994	14	8	3	2	1
1995	38	23	6	8	1
1996	38	25	7	5	1
1997	43	24	11	3	5
1998	47	35	4	7	1
1999	43	26	6	9	2
2000	28	17	2	7	2
Total	251	158	39	41	13
%	100	63	16	16	5
average ± standard deviation			1.31 ± 0.57	1.32 ± 0.65	1.38 ± 0.65 + 1.31 ± 0.63

the case wherein a group consisted of only one pair, the average group size of the Rufous Vanga is 3.5 individuals (maximum 6 individuals). This group size was recorded during the breeding period, and we do not have detailed records of the non-breeding period. However, according to the observations of Yamagishi and Masahiko Nakamura in August 1990, Rufous Vangas lived in small groups. We therefore infer that this species lives throughout the year as monogamous pairs or in a small group.

Territory

Whether a pair or a larger group, each Rufous Vanga group maintains a territory within which foraging and breeding takes place. Conflicts between neighboring groups or individuals at the territory boundary are common. On such occasions, the members of each group clearly separate and line up on tree branches at the border of their respective territories, each group facing the other. All group members, including the males, females, and helpers, participate and chirp loudly at the opposite group to proclaim the boundary of their territory. This back-and-forth chirping usually continues for an extended duration, but direct aggressive contact between members of opposing groups seldom occurs. However, on such occasions, the chirping sound consists of distinctive chirp patterns, different from the aforementioned three-note chirp, and they sometimes click their beaks, etc. Indeed, their chirping sounds are quite varied, and they emit distinctive chirps depending on the situation, such as during courtship or when intimidating predators. Their chirps appeared to differ in relation to the predator as well, e. g., they chirped differently according to whether a raptor or a snake was present. Moreover, males and females make different sounds, and the dominant (breeding) male chirps in a manner that was distinct from that of the helper males.

As we grew familiar with their various types of chirps, we were able to identify the situation they were in fairly well without actually seeing the birds. For example, we could gauge when they were engaged in a conflict against a neighboring group or when they were building a nest.

Figure 3-7 shows the composition of each group, and Figure 3-8 (folded page) shows the territorial distribution of these groups in Jardin A[4]. Although the latter figure might suggest that the entire forest area was densely packed with territories, the boundaries themselves were not so well defined. Within a territory, there were certain areas that were used frequently and other areas that were only used occasionally. Therefore, the actual territory distribution was not as clearly defined as that depicted in the figure. However, as mentioned above, most individuals within this population had already been banded with colored bands and had been individually identified. Therefore, apart from the groups shown, few other territories, if any, existed within the study area. The boundaries depicted in the figure should be interpreted as such.

Figure 3-8 indicates that while the location of the territories remains the same every year, careful examination reveals that their range is subtly modified. Consequently, we believe that maintaining the precise range of a territory is not too critical to successful breeding. New territories appear to be formed by pushing aside existing territories. For instance, in 1997, a new territory, Group-ZX, was established near the northwest intersection of trails. The territory distribution of the previous year indicates that the new territory used to be adjacent to HX, X2, BX, etc. This new territory consists of portions trimmed from already existing territories. Moreover, in this instance, both the pair in Group-ZX and the pairs that were pushed aside were able to breed. As long as they secure a certain amount of space, the Rufous Vangas appear to be flexible with regard to the size of their territory. Consequently, within this study area, it does not appear that helpers stay with the parents because the area is spatially saturated with territories and they have nowhere else to go. Figure 3-8 indicates that in the same study area, territories increase and decrease in number between years. This implies that there is no system whereby the number of helpers becoming independent corresponds to the number of territories that disappear.

The figure showing territory distribution indicates the location of each group shown in Figure 3-7 and the range of movement of the transferred individuals. Referring to these two figures will henceforth aid the reader in recognizing the location and history of any group mentioned in this work.

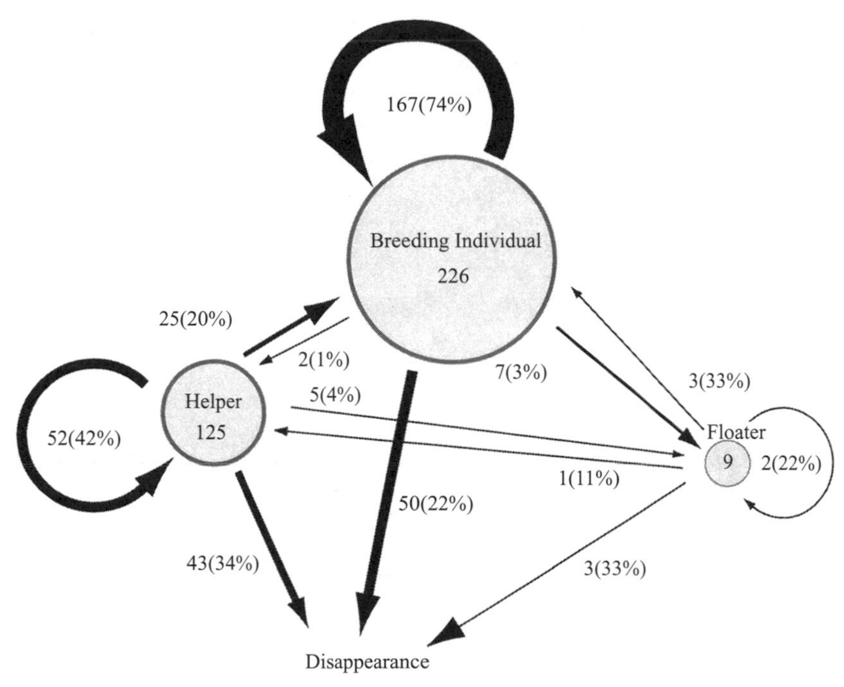

Fig. 3-9 Transition of male's status (total number from 1994 to 2000)
Arrows show the transition of status, and numbers attached to arrows indicate the total number of transfers during seven years. In general, males become breeding individuals after they experienced being helpers, and floaters are an exceptional case. The category of Disappearance includes individuals which transferred a far distance, died, or of which status is unknown (from Asai et al. 2001[4]).

The lives of males and females

Earlier in this chapter, 10 years in the life of the "pink-stripe banded male" were reviewed. Let us now focus on the course of life led by its male offspring. Figure 3-9 was derived, like other figures, on the basis of the data represented in Figure 3-7. Firstly, similar to the trend noted earlier, almost all yearling males became helpers. Of helper males 42% continued as helpers, and 34% disappeared from the study area. This 34% includes individuals that died and those that relocated outside the research area. Twenty percent became breeding males and 4% became floaters. Among individuals that achieved breeding male status, 74% became breeding males the following year as well, 22% disappeared from the study area, 3% became floaters, and 1% returned to being helpers.

It is far more difficult to track the life history of females since most of them leave their natal territory and move outside the study area. However, we do know that they become breeding females at age one, and it appears that they subsequently continue to be breeding females unless something unusual occurs.

Thus far, this chapter has described our research methods and has provided an overview of the manner in which our study progressed. A familiarity with these details is advisable as it will help the reader fully apprehend the subsequent detailed research findings regarding the social organization of the Rufous Vanga.

References

1 Langrand, O. (1990) *Guide to the Birds of Madagascar*. Yale University Press, New Haven & London.

2 Yamagishi, S. & Eguchi K. (1996) Comparative foraging ecology of Madagascar vangids (Vangidae). *Ibis*, 138: 283-290.

3 Yamagishi, S., Asai, S., Eguchi, K. & Wada, M. (2002) Spotted-throat individuals of Rufous Vanga *Schetba rufa* are yearling males and presumably sterile. *Ornithological Science*, 1: 95-99.

4 Asai, S., Yamagishi, S. & Eguchi, K. (2001) History of group compositions of the rufous vanga *Schetba rufa* at Ampijoroa in northwestern Madagascar. *Memoirs of the Faculty of Science, Kyoto University (Series of Biology)*, 17: 33-40.

The role of the Rufous Vanga in a mixed-species flock of birds

Teruaki Hino

A perfect theater for watching a mixed-species flock of birds

Most species of birds that inhabit temperate-zone forests, such as in Japan, throughout the year, form groups and forage together after the major event of breeding from spring to summer. A close examination reveals that these groups often consist of several different species of birds. Such a group is called a mixed-species flock, and in Japan such flocks primarily comprise species of tits and chikadees, typically Long-tailed Tits *Aegithalos caudatus*, Great Tits *Parus major*, Willow Tits *P. montanus*, and Coal Tits *Parus ater*. At times, these flocks also include woodpeckers.

As a graduate student of Hokkaido University, my research was directed toward understanding mixed-species flocks. My objective was to clarify the behavioral interactions between flock members at the individual level, such as foraging and attacking, and the structural organization at the community level, i.e., within a gathering of several species. As part of my efforts to study the inter-individual differences between flock members, I banded the legs of each bird with a unique combination of colored rings. At that time, few studies had examined bird flocks from this perspective. Hence, I believed that I was undertaking a potentially interesting and important study. Unfortunately, not everything proceeded according to plan. For example, under the natural conditions in the forest, I was unable to track and monitor each individual's behavior and the flock structure. To evaluate the adaptive significance of a mixed-species flock, it is necessary to compare each individual's foraging efficiency (food intake quantity per unit time) in relation to flock size and species composition. However, in the forests of Hokkaido, it was difficult to conduct detailed observations because the birds foraged mostly in the canopy at heights above 15 m. Moreover, the small body size of the main subject species, the tits and chickadees, whose body size is less than 15 cm, caused further complications in our attempts at detailed observation. Consequently, I was unable to obtain satisfactory data even when I used binoculars of 10 × magnification. Therefore, as a final resort, foraging and behavioral observations had to be conducted at artificial feeding stations[1].

Despite these drawbacks, I eventually completed my thesis, and in 1992, I secured a job at the Forestry and Forest Products Research Institute. It was during the autumn of that year that Dr Yamagishi invited me to join his planned research project in Madagascar, for which he was applying to the Japanese Ministry of Education, Culture, Sports, Science and Technology for funding. Since I was eager to begin work on a novel and interesting research theme at that time, I accepted his invitation without any hesitation. Dr Yamagishi requested that I study "The effects of inter-specific sociological relationships within mixed-species flocks on the society of the Rufous Vanga." My first thought was: "This sounds difficult!" Nevertheless, I was overjoyed at the prospect of conducting research in Madagascar, the "wonderland of biodiversity." Unfortunately, the application for funding was rejected that year; thus, it was not until October 8, 1994 that I visited the Ampijoroa Research Station for the first time.

On the day after my arrival, the first thing I did was to visit Jardin A. The forest there was relatively sparse, similar to an early secondary-growth forest in Japan, and most of the trees were merely 12 m in height (Fig 2-6, Chapter 2)[2]. After taking only a few steps into the forest, it quickly became apparent that it was home to an amazing diversity of birds. Moreover, since most of these birds were over 16 cm in body size, behavioral observations were possible even without binoculars. During the course of my research on mixed-species flocks in Hokkaido, I was unable to conduct satisfactory research; however, it seemed possible for me to achieve my goals here. Jardin A, which is French for "Garden A," appeared to be the perfect theater in which to enjoy the performances of a cast of birds. I was excited at the prospect and potential of conducting research there.

Dr Eguchi, the acting team leader in Madagascar, informed us of our schedule. During the entire morning, team members would participate in research on the Rufous Vanga, while in the late afternoon/evening, researchers could work on their individual study themes. The first week of collaborative research was spent netting Rufous Vangas in order to leg-band them to facilitate individual identification, and draw blood samples for future DNA studies. Near the mist net, we played a recording of the distinctive three-syllable chirp of the Rufous Vanga (Fig 3-19). We then waited as the birds approached the source of the sound to defend their territory from the perceived intruders and flew unawares into the mist net. However, during this exercise, an unexpected event occurred—the first bird to approach the vicinity of the recorder was not a Rufous Vanga but a Madagascar Paradise Flycatcher! Subsequently, not only Rufous Vangas but several species of birds gathered round. It seemed as if the entire cast of members of the living theater had responded to a roll call. Why did this happen? This mystery was solved later during the course of our research on mixed-species flocks; however, at that moment, it came as a refreshing surprise.

Madagascar lies in the southern hemisphere, and October heralds the onset of the rainy season, during which most birds in this country begin breeding. As is observed in Japan, several breeding birds in Madagascar defend their territories by vocal threats (chirping). During the evenings of the first week, I undertook a census of the number of birds through territory mapping. This exercise also aided me in learning to identify the species found in that area. To estimate the number and location of the territories, I systematically traversed the entire research area and plotted the points where birds chirped and conflict occurred, the direction in which they moved, etc., on a map. During the course of my later research on mixed-species flocks, I continued this plotting method and documented the fact that 124 pairs of birds, comprising 29 species (excluding raptors), were breeding in the forest of Jardin A[2]. In addition, I documented the fact that 14 species, i.e., half the total number of species living there, participate in mixed-species flocks (Fig 4-1).

That year, research on mixed-species flocks was conducted for approximately one month from mid-October until mid-November. Whenever I encountered a bird within the research area, I first noted whether the bird was foraging in solitary or with a group of conspecifics or in a mixed-species flock. If part of a flock, I also noted the number of individuals in the flock, and if part of a mixed-species flock, I noted the species composition of the flock. Thereafter, I observed a focal individual for two minutes, recording foraging method and other behaviors. In cases where aggressive and submissive behaviors were observed, I recorded the species of the interactors. Since our project was subject to a time limitation, I was forced to abandon my attempts to identify all individuals by color banding them. However, I put in my best efforts toward recording the gender, age, morphology, and dominant or subordinate social status of as many individuals within the flock as possible. Gathering such comprehensive data on a solitary individual was relatively easy, but when recording data from mixed-species flocks, I was extremely busy and had no respite. To simplify data collection in such situations, data was recorded verbally via a tape recorder. Later, after the evening meal or on our weekly holiday, the results were transcribed by hand onto data sheets. I attempted to collect data on the maximum possible number of individuals; hence, during observations, I kept

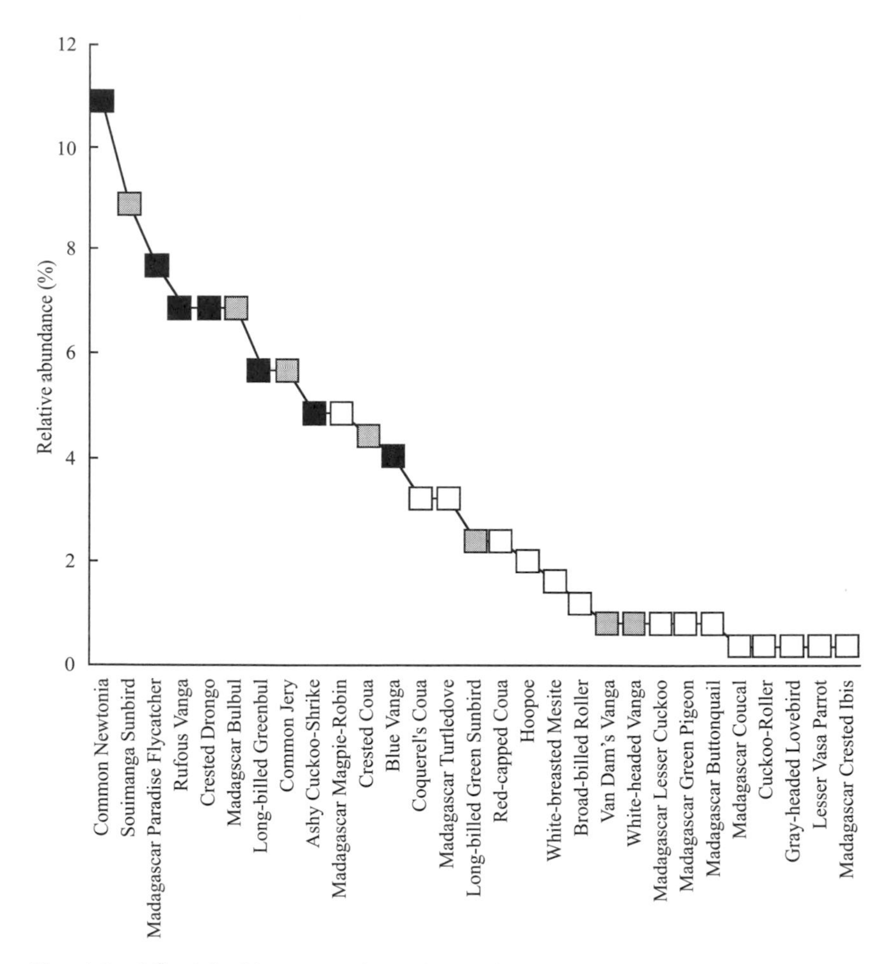

Fig. 4-1. Mixed-flocking propensity and the relative abundance of species arranged in decreasing order of abundance. Black >40%, gray 20-40% and white <20% particiation mixed-speces flocks. (From Hino 2002)

frantically recording my observations onto the recorder for an uninterrupted hour. Despite this effort, on one occasion, I was horrified to realize that the recorder battery was dead and that none of the data I had input during the past hour had been recorded. However, the feeling of satisfaction and achievement that I experienced after successfully accumulating my target data can never be traded. After returning to Japan, I spent one week inputting these data into a computer spreadsheet, and later, I performed an analysis.

Give-and-take relationships in mixed-species flocks

As a result of the 1994 research, the main characters in this mixed-species flock theater troupe were revealed: the short-tempered Rufous Vanga, who is the leading character of this book; the Blue Vanga *Leptopterus* (*Cyanolanius*) *madagascarinus*, who belongs to the same family as the Rufous Vanga and has a beautiful blue-plumaged body; the Crested Drongo *Dicrurus forficatus*, who appears rather cunning and has a head adorned with an elegant crown of feathers; the Madagascar Paradise Flycatcher, a species wherein the males have distinctive tail feathers; the small, quick-moving Common Newtonia *Newtonia brunneicauda*; the Long-billed Greenbul, which resembles the Japanese Bush Warbler *Cettia diphone*; the Ashy Cuckoo-Shrike *Coracina cineria*, which has an unsophisticated demeanor; and so on. All of these were birds with a great deal of character (Plate I-12).

For approximately half the number of the occasions on which I observed them, I found each of these seven species to be in the company of other species (Fig 4-2). Other than that, they foraged alone or with conspecifics (in several cases with their breeding partner). This behavior differs from that witnessed among birds inhabiting temperate forests, such as in Japan, where such behavior is observed only during the non-breeding season. It is not witnessed during the breeding period because the birds are busy building nests, brooding their eggs and young, and feeding the chicks. In Madagascar, however, despite breeding activities, these seven species were spending half of each day in the mixed-species flock. From books and papers[3,4] on this subject, I was already aware that in the tropical forests, mixed-species flocks are present throughout the year. However, after a firsthand experience of this phenomenon, I was eager to learn the reason for this occurrence.

Apart from the singular appearance of each leading actor, the foraging performance (height, substrate, technique) at which each was most proficient was also unique. However, a comparison of the foraging performance used by the various species when foraging within a mixed-species flock and otherwise revealed that six of the seven species used different performance (Figs. 4-3, 4-4, 4-5)[5]. The Crested Drongo exhibited the greatest difference. When foraging with conspecifics or solitarily, the Crested Drongo perched upon branches at a height of >6 m and flew and caught small insects such as bees and flies. However, after joining a mixed-species flock, this bird descended to a height of <6 m and began to forage on large arthropods, such as caterpillars, centipedes, spiders, and grasshoppers, which live on the branches, in the foliage, and on the ground. Although not as conspicuous as the change in foraging performance observed in

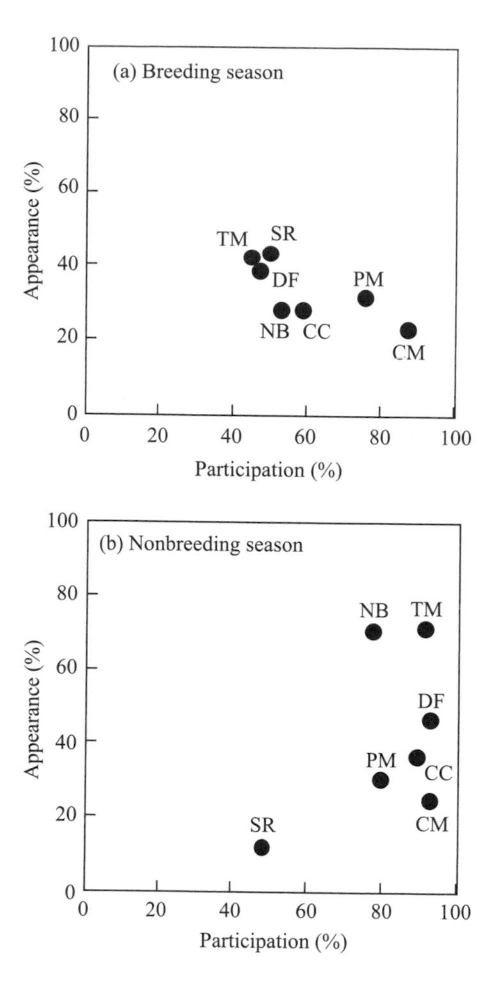

Fig. 4-2. The percentages of participation (=Number of observations in flocks/ Total number of observations) and appearance (=Number of presence in flocks/ Total number of flocks) in mixed-species flocks by the seven regular flock-participating species in the breeding season (a) and the nonbreeding season (b). RV=Rufous Vanga, NB=Common Newtonia, DR=Crested Drongo, PF=Paradise Flycatcher, BV=Blue Vanga, LG=Long-billed Greenbul, CS=Ashy Cuckoo Shrike.

the case of the Crested Drongo, the Madagascar Paradise Flycatcher also altered its behavior when it joined the mixed-species flock. Its principal foraging method changed from sallying to catch insects on the wing to hovering to catch insects living in the foliage. This change was particularly evident in those males that had long tails[6]. They also began foraging in bushes and on the ground surface, which they never did when foraging alone or with conspecifics.

The Blue Vanga frequently fed while hanging upside down on a branch, whereas the Common Newtonia picked its food items from branches and foliage through equal use of a sallying metrod, a hovering-by-fluttering method and a gleaning method. However, when in mixed-species flocks, both these species foraged at a lower canopy level than that at which they foraged when in solitary or in the company of conspecifics. The Ashy Cuckoo-Shrike possessed the most diverse foraging repertoire among the seven main species, and the Long-billed Greenbul primarily fed by probing at the lower strata of <6 m. However, when in mixed-species flocks, these two species foraged at various levels of the forest strata. We analyzed our data on foraging behavior by means of an overlap index. This analysis revealed that when a species joined a mixed-species flock, the species-specific differences in foraging techiques and the differences in height and substrate at which foraging was performed became less pronounced than when a species foraged solitarily or in conspecific groups[5].

We also examined the foraging speed (the frequency of foraging per unit time), which is an indicator of whether or not birds are foraging efficiently. We compared the effects that foraging solitarily, in a conspecific flock, and in a mixed-species flock had on foraging speed. Our analysis revealed that whether they foraged solitarily or in a conspecific flock, there was no difference in the foraging speed of any of the seven species. However, among five species—the Crested Drongo, the Madagascar Paradise Flycatcher, the Common Newtonia, the Ashy Cuckoo-Shrike, and the Long-billed Greenbul—it was found that the foraging speed increased significantly when foraging was performed in mixed-species flocks (Fig 4-6). This implies that for these five species, the mixed-species flock is a beneficial aggregation in terms of improving their foraging performance, i.e., foraging optimally.

What then is the relationship between the similarity in the foraging site and an increase in foraging efficiency? Four possibilities could be suggested. The first is the effect of "social learning"[7]. By observing the behavior of other birds foraging in the same flock and subsequently foraging in the same place using an identical method, an individual may successfully find food and thereby increase its foraging efficiency. As mentioned above, each species' foraging site and method differs according to what it is most proficient at. Therefore, an individual's advantage of social learning is considered to be greater when it forages with heterospecifics that exhibit different foraging methods and utilize different sites than when it forages with conspecifics that forage at the same site and employ the same method. We could consider that all the five species whose foraging speeds increased in mixed-species flocks derived this advantage. This give-and-take relationship results in a bilateral benefit due to the interaction

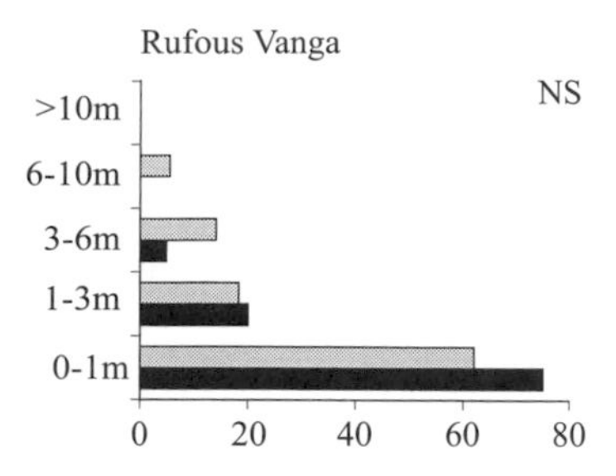

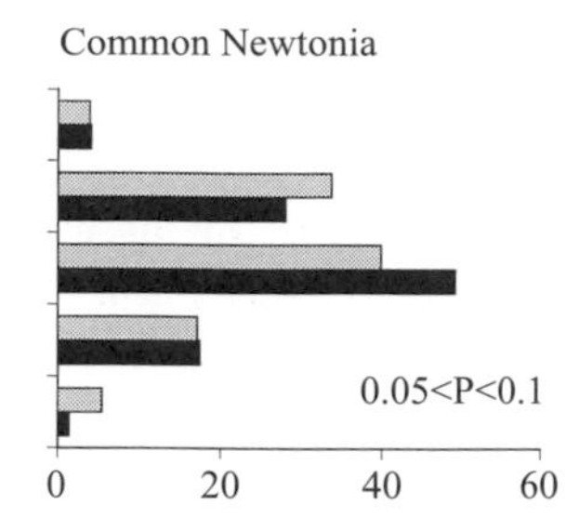

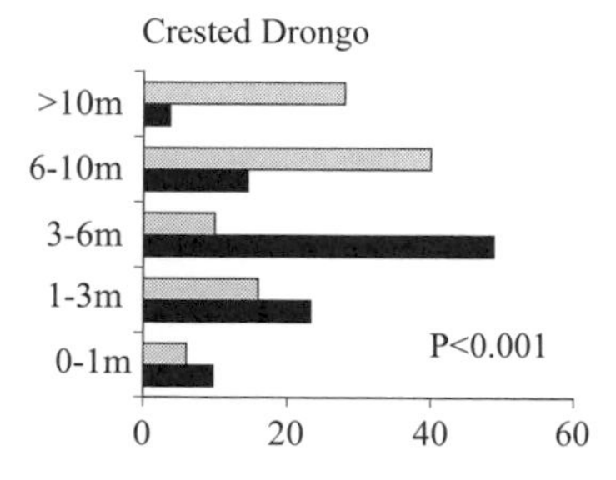

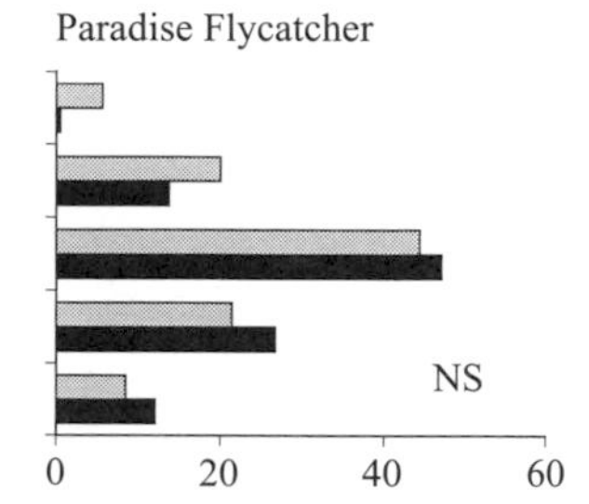

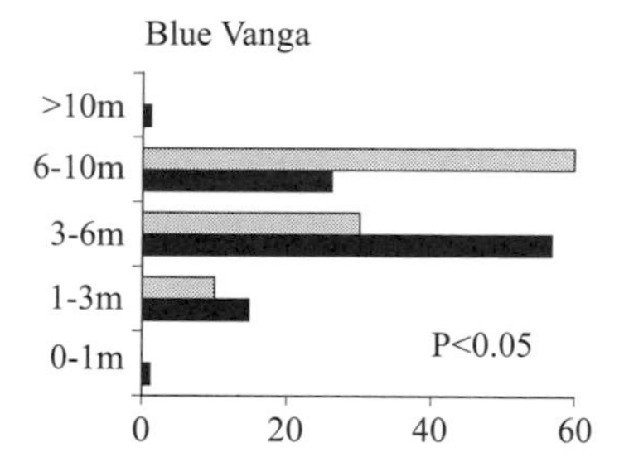

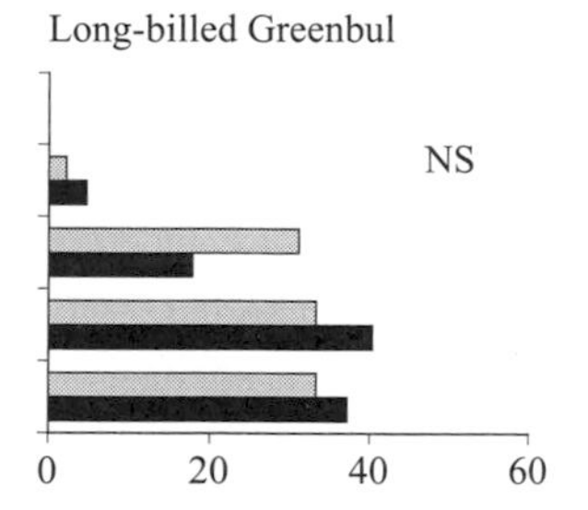

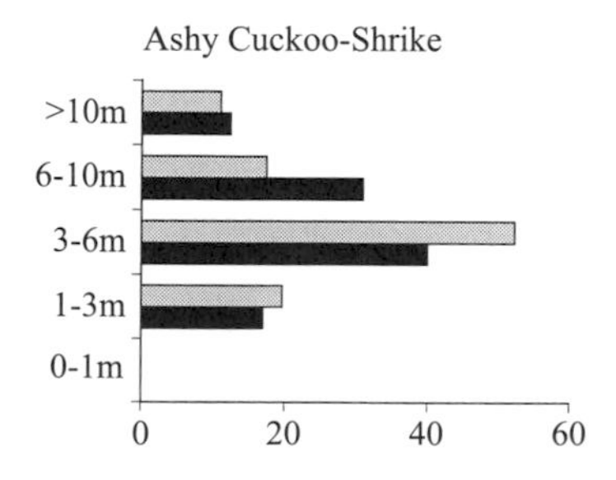

Fig. 4-3. Percentage use of height categories for prey-capture by the seven regular flock-participating species when solitary or in conspecific flocks (gray), or heterospecific flocks (black). Figures show the results of χ^2 test, NS P>0.1. (From Hino 1998)

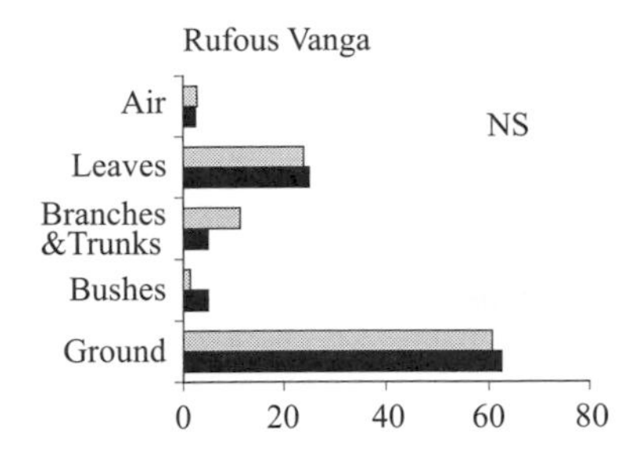

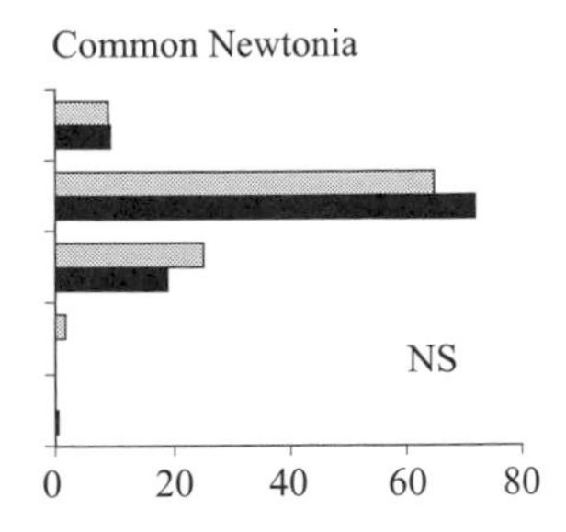

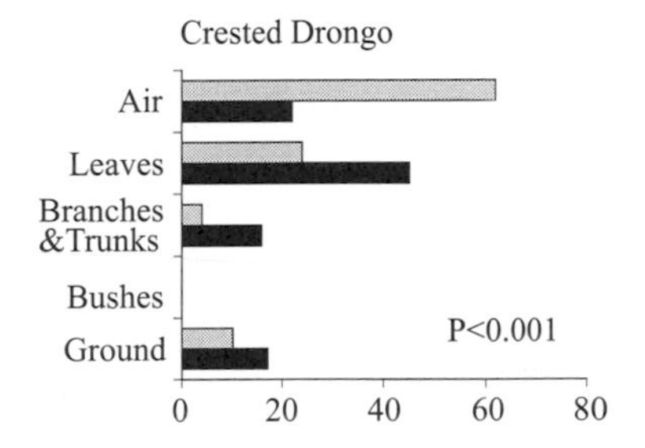

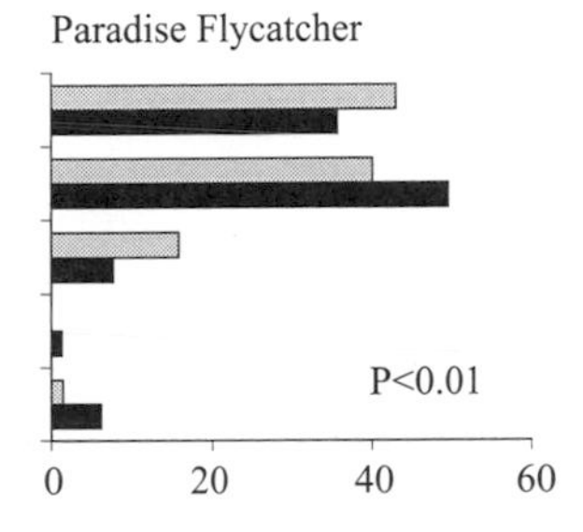

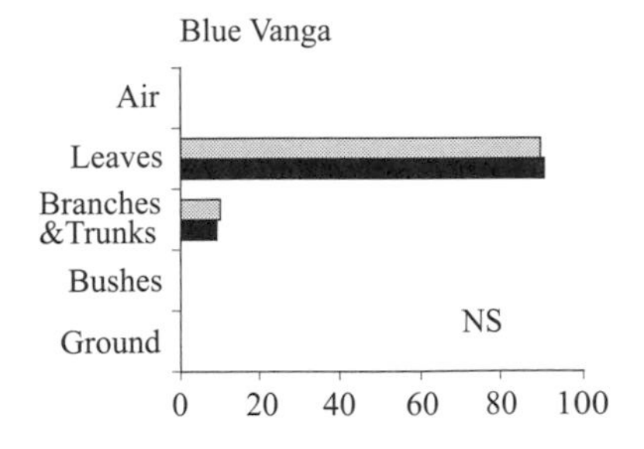

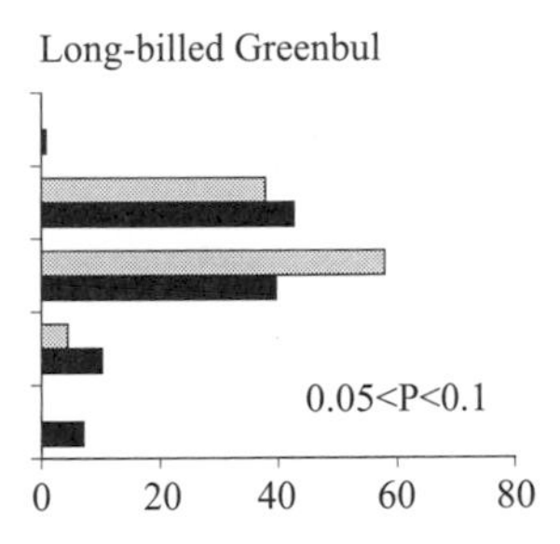

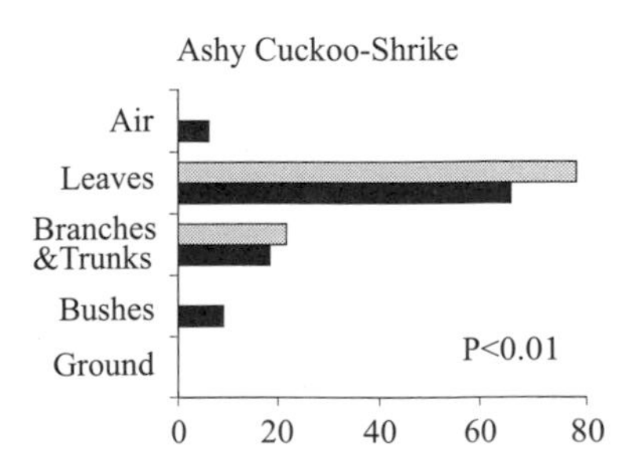

Fig. 4-4. Percentage use of substrate categories for prey-capture by the seven regular flock-participating species when solitary or in conspecific flocks (gray), or in hetero-specific flocks (black). Figures show the results of χ^2- test, NS P>0.1. (From Hino 1998)

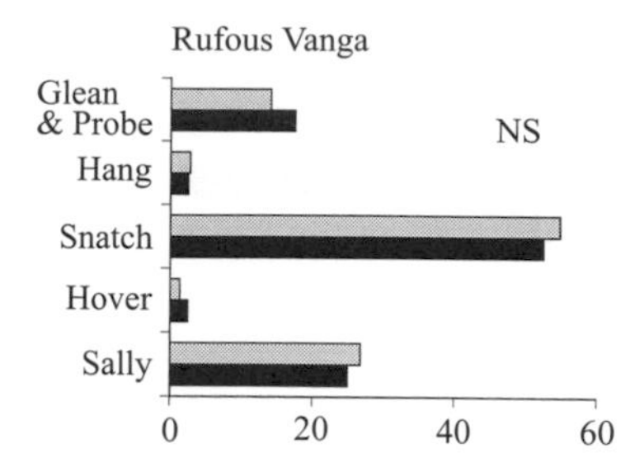

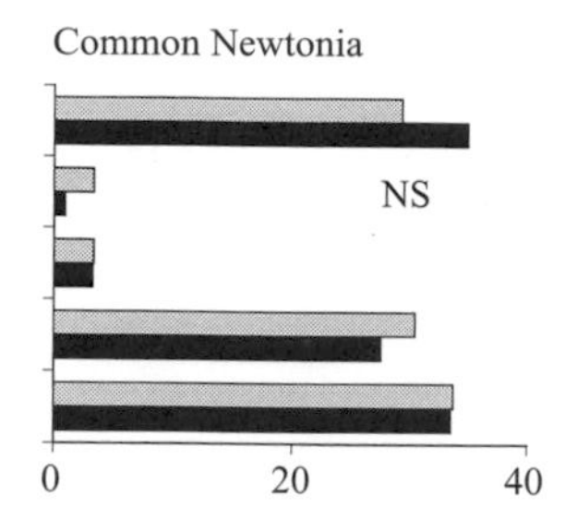

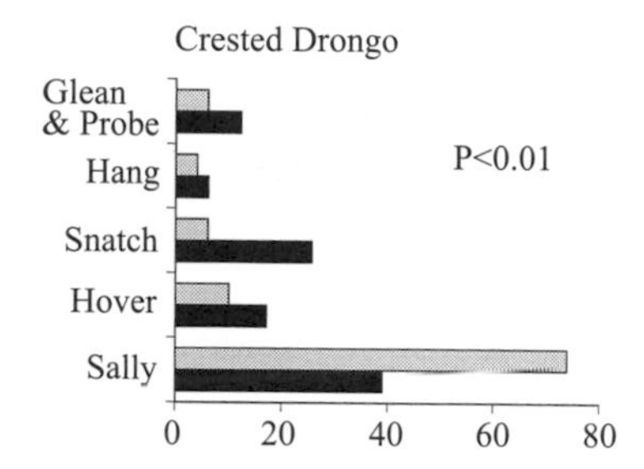

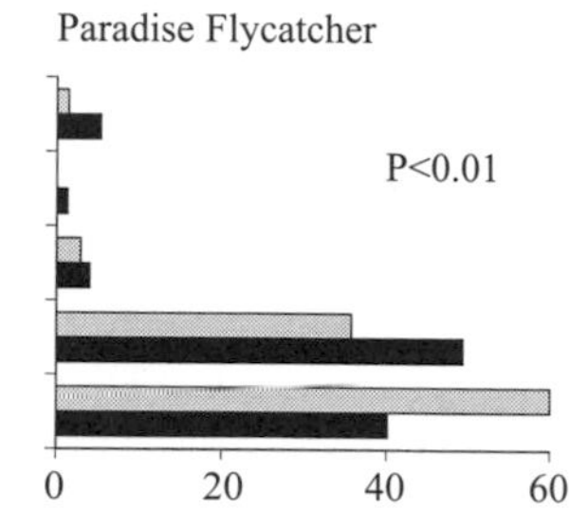

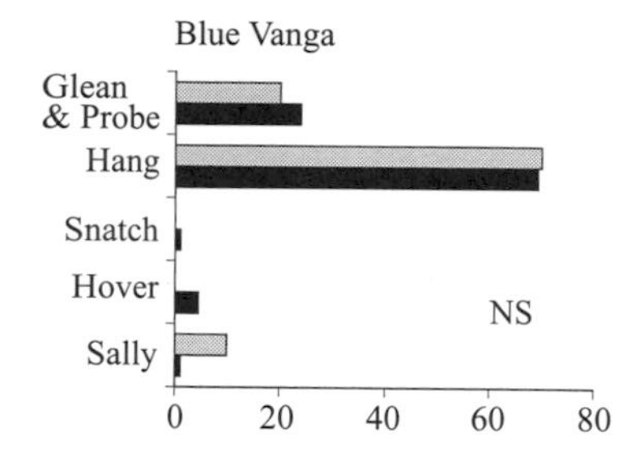

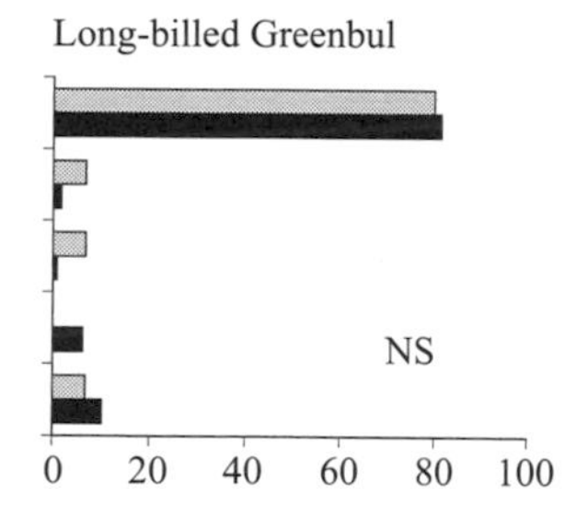

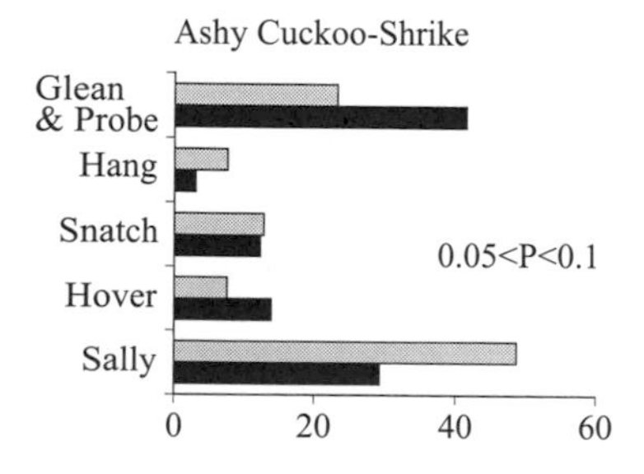

Fig. 4-5. Percentage use of techniques for prey-capture by the seven regular flock-participating species when solitary or in conspecific flocks (gray), or in heterospecific flocks (black). Figures show the results of χ^2- test, NS $P>0.1$.(From Hino 1998)

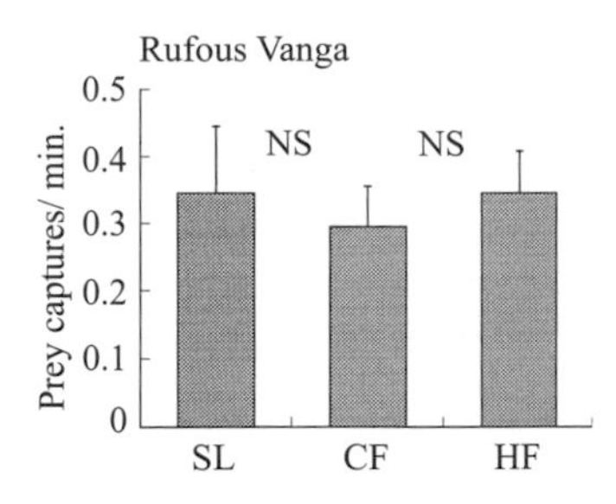

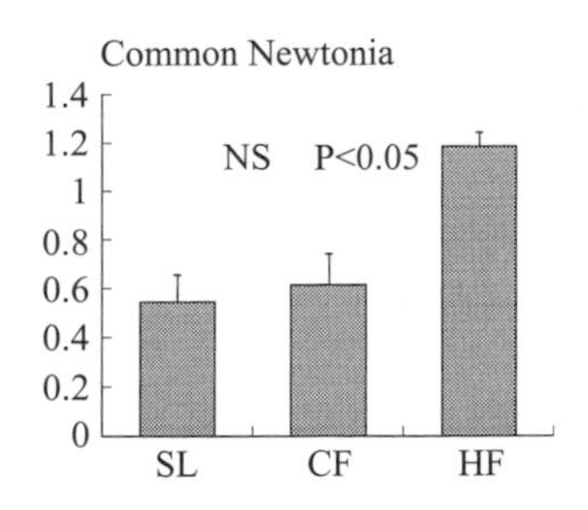

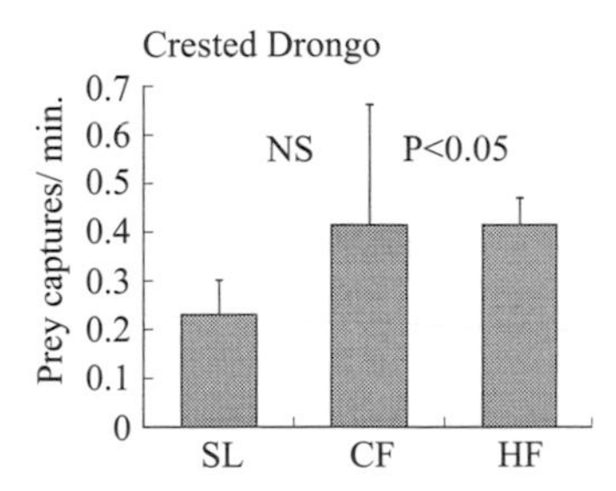

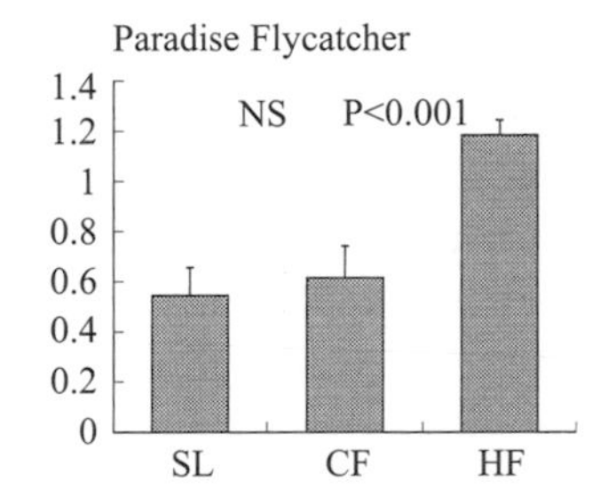

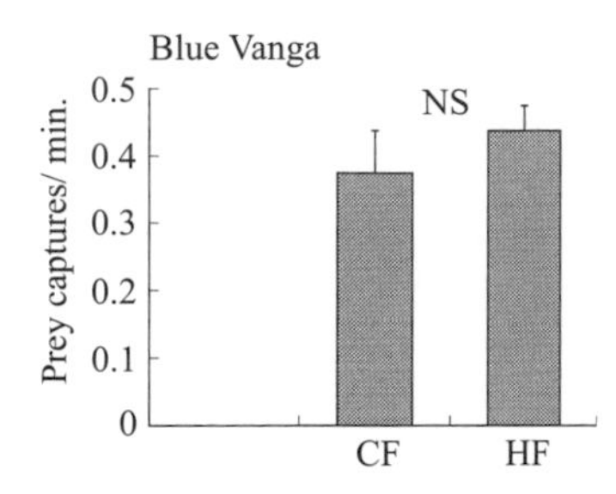

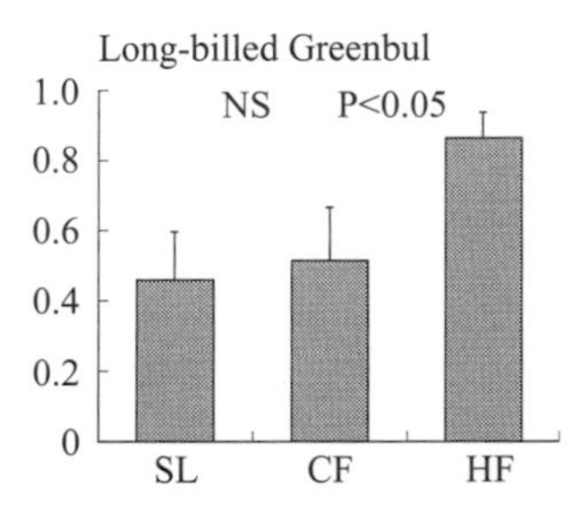

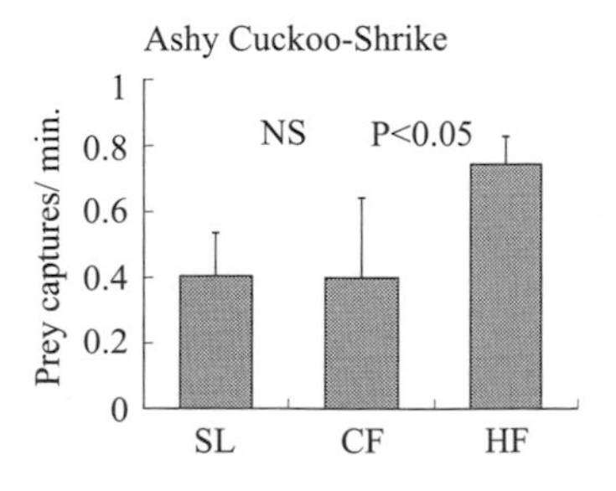

Fig. 4-6. Average feeding rates by the seven regular flock-participating species when solitary (SL), and when in conspecific (CF) or heterospecific flocks (HF). Figures show the results of U - test between SL and CF (left) and those between SL+CF and HF, NS $P>0.1$. (From Hino 1998)

among mixed-flock members, and it is termed "mutualism." The advantage of "foraging time increase" can similarly be explained in terms of mutualism[8]. The advantage of foraging in mixed-species flocks is gained through a reduction in the time spent in anti-predator vigilance; this allows for more foraging time. Furthermore, since it is believed that each species has a slightly different anti-predator vigilance range, the advantage could be larger in a mixed-species flock than in a conspecific flock.

Other than these mutualistic relations, another advantageous relationship possible between species in mixed-species flocks is a "commensal relationship," whereby one of the interacting species derives benefit while the other is not negatively affected. The movement of several individuals in a flock through the canopy disturbs numerous insects, which then emerge from the branches and foliage[9]. For the Madagascar Paradise Flycatcher, which feeds primarily on such food items, this "beating" effect creates an ideal situation. This is an advantage that they cannot derive through membership of a conspecific flock.

A certain type of behavior occasionally exhibited by the Crested Drongo also increased the foraging efficiency of this species. This behavior is termed "kleptoparasitism"[7], which implies that an individual does not actively search for food but instead, waits for other species to locate and capture food items, and then chases them away and steals their food[10]. This behavior, whereby they vigilantly watch for an opportunity to steal, befits their cunning appearance. In this relationship, the loss is borne by the opponent, whose food is stolen by the Crested Drongo. However, in mixed-species flocks observed in the temperate zone, such as in Japan, such behaviors occur frequently and among several species. I was intrigued by the difference observed in the frequency of such behavior between the mixed-species flocks of temperate forests and those of Madagascar. Later in this chapter, I will delve further into this issue.

The reliable Rufous Vanga

What is the role played by the Rufous Vanga, the leading character of this book, in mixed-species flocks? As opposed to the other species, the Rufous Vanga exhibited no change in foraging behavior when in mixed-species flocks. Instead, individuals of this species continued to forage by snatching, targeting large-size arthropods, such as mole crickets, spiders, centipedes, etc., and small lizards and chameleons, all of which live on or near the ground. Membership in a mixed-species flock did not result in any increase in foraging speed either. These facts indicate that this species "goes its own way" and is unaffected by the presence of other species. However, despite this fact, it was clear that the Rufous Vanga was the nucleus of mixed-species flocks in the forest of Jardin A during the breeding season.

A mixed-species flock is not a simple aggregation of varied species that have gathered together indefinitely. Rather, this aggregation should be considered the result of an interconnection, wherein one species follows another species in order to obtain some benefit. Therefore, both the "leader" and the "follower" play an important role in mixed-species flocks; the former leads other species while the latter follows other individuals. By recording any prominent "following behaviors" whenever we observed them, we were able to identify the role of each species. This analysis revealed that the Rufous Vanga and the Common Newtonia were leaders while the Crested Drongo and the Madagascar Paradise Flycatcher were followers. The other three species occupied indeterminate roles, assuming the role of leaders on certain occasions and that of followers on others (Table 4-1).

Among all the members, the Madagascar Paradise Flycatcher and the Crested Drongo derived the most benefit from their participation in the mixed-species flock. Since the Madagascar Paradise Flycatcher benefits via the beating effects of other species and the Crested Drongo benefits by social learning and kleptoparasitism, it is easy to understand why they play the role of followers within the flock. The Madagascar Paradise Flycatcher often followed the Common Newtonia, which has a similar foraging site and method, rather than the Rufous Vanga. In reality, mixed-species flocks consisting of only these two species were often observed. Moreover, data indicates that the greater the number of Common Newtonia foraging together in the flock, the greater the increase in the foraging efficiency of the Madagascar Paradise Flycatcher[6]. However, when the Rufous Vanga was present in a mixed-species flock, the Madagascar Paradise Flycatcher flew just above the ground surface and foraged there. The Madagascar Paradise Flycatcher is the only species whose foraging performance altered with respect to the presence or absence of the Rufous Vanga. The Crested Drongo too was not completely indifferent to the foraging method of the Rufous Vanga. I often witnessed members of this species observing and following Rufous Vangas that were picking insects from the ground surface. It appeared as if the Crested Drongo was trying to poach the food that the Rufous Vanga failed to capture. However, I was unable to confirm whether this bird was successful in its poaching attempts. This is possibly because the aggressive dominance of the Rufous Vanga in this relationship ensures that the Crested Drongo maintains a certain distance from it since venturing any closer might trigger an attack.

The two follower species often forage solitarily. In contrast, it was observed that the leader species, the Common Newtonia and the Rufous Vanga, foraged

Table 4-1 Leader-Follower relationship among the seven regular flock-participating species during the (a) breeding and (b) nonbreeding seasons. The figures show the incidence of following by a species in the left column toward a species in the top row.

(a) Breeding Season

Species leading	Rufous Vanga	Common Newtonia	Ashy Cuckoo-Shrike	Blue Vanga	Long-billed Greenbul	Paradise Flycatcher	Crested Drongo	Other Species
Species following								
Rufous Vanga	*				1			1
Common Newtonia		*		1				
Ashy Cuckoo-Shrike	4		*					
Blue Vanga	3	1	2	*				
Long-billed Greenbul	6	1	2	1	*			1
Paradise Flycatcher	4	8		2	2	*	1	3
Crested Drongo	5		3	2	2	2	*	3
Other Species		1	1			1	1	*
Total as leading	22	11	8	6	5	3	2	
Total as following	2	1	4	6	11	20	17	
% as leading	92	92	67	50	31	13	11	

(b) Non-breeding Season

Species leading	Rufous Vanga	Common Newtonia	Ashy Cuckoo-Shrike	Blue Vanga	Long-billed Greenbul	Paradise Flycatcher	Crested Drongo	Other Species
Species following								
Rufous Vanga	*							1
Common Newtonia		*				2		1
Ashy Cuckoo-Shrike	2	1	*	2	2		2	2
Blue Vanga	2		2	*	2			2
Long-billed Greenbul	2		1	1	*		1	
Paradise Flycatcher	3	34		2	3	*	0	6
Crested Drongo		2	8	6	12	1	*	7
Other Species	1	2						*
Total as leading	10	40	11	11	19	3	3	
Total as following	1	3	11	8	5	48	36	
% as leading	91	93	50	58	79	6	8	

together with more than three conspecifics even during the breeding period. As described in the other chapters, a group of Rufous Vangas is a breeding unit that consists of helper individuals; however, the helper system is alien to the Common Newtonia. Almost always, it was the Madagascar Paradise Flycatcher who followed the Common Newtonia, and by doing so, its foraging efficiency increased. However, though most of the other species followed the Rufous Vanga, their foraging efficiency did not increase. This was possibly due to the fact that the foraging sites of the Rufous Vanga are different from the usual foraging sites wherein the follower species are adept. In fact, the Long-billed Greenbul, which had strong ties with the Rufous Vanga, showed a decrease in foraging speed. Yet, this species chose the Rufous Vanga as a heterospecific

Table 4-2 Dominance relationship among the seven regular flock-participating species. The figures show the number of attacks by a species in the top row against a species in the left column. (Hino 1998)

Species attacking	Rufous Vanga	Crested Drongo	Paradise Flycatch-er	Ashy Cuckoo-Shrike	Blue Vanga	Long-billed Greenbul	Common Newtonia	Other Species
Species attacked								
Rufous Vanga	*			1				
Crested Drongo	4	*			1			2
Paradise Flycatcher	3	4	*				2	
Ashy Cuckoo-Shrike		3		*				1
Blue Vanga		4	1	1	*			1
Long-billed Greenbul	4	3				*		
Common Newtonia	1	1	9	1	2	2	*	2
Other Species	3	1					1	*
Total as attacking	15	16	11	2	3	2	3	
Total as attacked	1	7	9	4	7	7	18	
% as attacking	94	70	55	33	30	22	14	

foraging partner during the breeding season. The fact that a recording of the chirping sound of the Rufous Vanga was sufficient to draw several birds toward the sound and trap them is evidence of the considerable attractive power of the Rufous Vanga's chirp. In fact, although the appearance rate of the Rufous Vanga in mixed-species flocks was slightly lower than 50%, it was the highest recorded value, and it exceeded the appearance rates of the two follower species—the Madagascar Paradise Flycatcher and the Crested Drongo (Fig 4-2).

How can we interpret the performances of the cast in the mixed-species flock ensemble of Jardin A, wherein the leading actor is the Rufous Vanga? Also in the mixed-species flock of tits and chickadees in Hokkaido where I conducted my graduate student research, the roles of leader and follower are determined depending upon species. During the non-breeding period, the Long-tailed Tit forms a large conspecific flock and is always a leader, whereas the Great Tit, in whose case the interconnection between conspecifics is weak, is a follower. As observed among the flock in Jardin A, the number of individuals that forage together is one of the factors characterizing leader and follower status. However, the relationships for social order were different. Within these mixed-species flocks of tits and chickadees, species with a smaller body size and weaker fighting ability were ranked lower; these were followed by higher-ranked species of larger body size and stronger fighting ability. Therefore, the social order from lower to higher, for example, Long-tailed Tit – Coal Tit – Willow Tit – Great Tit, is precisely the relationship between leader and follower. This situation occurs because the higher-ranked species are able to approach lower-ranked species from the rear, drive them away, and poach their food[1]. As noted earlier, in the Jardin A forest, such kleptoparasitic behavior was observed only in the case of the Crested Drongo.

Such a relationship is inapplicable to the mixed-species flock of the Jardin A forest because the Rufous Vanga is the highest-ranked species within it (Fig 4-2). However, as was observed in the case of the Great Tit in Japan, Rufous Vangas did not attack other species among the flock to steal food from them. Instead, they invested their own efforts in searching for and capturing food. Observations of the Rufous Vangas when they attacked other flock members suggested that these attacks were primarily in the form of behavioral gestures that served to intimidate other birds and that these attacks resulted from the Rufous Vanga's irritation at being followed. However, if the source of the irritation was a raptor, the story was altogether different! Several raptors that feed on small birds, such as the Madagascar Sparrowhawk *Accipiter madagascariensis*, and Frances's Sparrowhawk, are found in the forest of Jardin A[2]. Whenever these predators appeared in the vicinity, a group of Rufous Vangas would inevitably be the first to sound the alarm by making a loud chirping noise. Occasionally, they bravely mobbed the predator and drove it away. The Brown Lemur *Eulemur fulvus*, which is believed to eat eggs and chicks in nests, and a kind of snake whose body length can measure up to 150 cm, were also targets of this behavior (Fig 5-10, Plate II-2). Faced by a group of Rufous Vangas with their sharp hooked beaks, retreat is perhaps the only alternative available to the predator.

These defensive attributes of the Rufous Vanga aimed at fending off predators are considered to be attractive to other members of a mixed-species flock; therefore, they positively follow them, i.e., "If seeking shelter, look for a big tree" (Plate I-13). The Rufous Vanga neither receives any benefits nor endures a loss through this relationship; thus, the relationship is commensal. The presence of the Rufous Vanga ensures the safety of other members and permits them, in turn, to gain an advantage in terms of foraging efficiency through social learning, etc. Interestingly, only one species, the Blue Vanga, exhibited no increase in foraging speed when participating in a mixed-species flock; however, some change in foraging site was evident. The foraging method of the Blue Vanga includes hanging upside down to pick food items from the underside of foliage. While in this position, this bird appears entirely defenseless against aerial predators such as raptors. Therefore, although they belong to the same family as the Rufous Vanga, ensuring safety by foraging together with short-tempered colleagues is perhaps an important strategy for this rather gentle species.

The Rufous Vanga that goes its own way

The following year, on August 21, 1995, I revisited the Ampijoroa Research Station. It still being the dry season, the birds living there were scheduled to begin their breeding activities only two months later. The research objectives of that year were to investigate the birds' activities during the non-breeding season and to compare this data with that collected during the breeding season of the previous year. Although I had not seen the birds of Jardin A for 10 months, the leading actors and other cast members were unchanged. I was thrilled to see the appearance and behaviors of these actors, which have varied characteristics, and when I encountered several Rufous Vangas and Madagascar Paradise Flycatchers bearing the colored bands we had placed on them during the previous year, I felt as though I had renewed my acquaintance with old friends.

The frequency with which I encountered mixed-species flocks was clearly higher than that during the breeding period. This was because the rate of participation of component members in the mixed-species flock had increased. In six of the seven leading character species, the participation rate was 80% or more, implying that almost the entire day was spent in the company of other species (Fig 4-2). In Japan as well, differences in mixed-flock membership are observed between the breeding and non-breeding periods; this situation was thus expected. More impressively, a great change was underway in the roles of certain members of the mixed-species flock theater group. Notably, the Rufous Vanga, which formed the nucleus and played a reliable role in the breeding season during the previous year, now had an inconspicuous presence within the group. In this species, the participation rate in mixed-species flocks was only 50%, the same as during the breeding period. Since the frequency at which other birds were observed in the mixed-species flock had increased, the relative presence of the Rufous Vanga was diluted, rendering it relatively inconspicuous. In reality, the appearance rate of the Rufous Vanga had decreased to only slightly above 10%, as compared to a value of 50% noted during the breeding season-the highest rate among flock members at that time. The Rufous Vanga seldom followed other birds, and so its role as leader remained unchanged. However, the frequency at which they were followed by other birds decreased. As usual, the Madagascar Paradise Flycatcher and the Crested Drongo were followers; however, on this occasion, the former followed only the Common Newtonia, while the latter followed the Blue Vanga, the Ashy Cuckoo-Shrike, and the Long-billed Greenbul (Table 4-1).

How did this change occur? The Rufous Vanga is a bird that goes its own way. Although there was a marginal difference during the non-breeding period since they hunted to a greater extent on the ground surface, their indifference to the presence of other species was unaltered from that observed during the breeding period. Accordingly, the degree to which the Rufous Vanga depended on the mixed-species flock also remained unchanged. The requirements of birds foraging around the Rufous Vanga in the mixed-species flock appear to have differed according to the season. It is therefore appropriate to suppose that their degree of dependence on the Rufous Vanga also differed. As mentioned above, during the breeding period, birds foraging around the Rufous Vanga relied upon its anti-predator defensive abilities. Did this advantage lessen during the non-breeding period? This is highly probable. As compared to the breeding period, when they have to protect their own chicks, the need for the Rufous Vanga to sound the alarm and to attack potential predators must have decreased significantly during the non-breeding period. In fact, my field notes contain no record of this bird having attacked any raptors or lemurs during this period—the behavior that took me by surprise during the previous year. During that period, I was under the impression that the Rufous Vanga was a short-tempered bird; hence, their apparent gentleness during this season belied my expectations.

Except for the Rufous Vanga, the number of conspecifics foraging together in all other bird species was higher in the non-breeding period than in the breeding period[5]. This is because during the breeding period, they forage individually or as pairs within their respective territories, whereas during the non-breeding period, the flock size is increased by the addition of individuals fledged during the previous year, immigrant individuals, and even individuals from neighboring territories. The large flocks of the Common Newtonia and the Long-billed Greenbul are perhaps examples of the latter, i.e., individuals from neighboring territories that join to form a flock. We infer this on the basis of our observations that both the flock size of each species and the average size of mixed-species flocks increased by 1. 5 times. The larger the flock, the greater the degree of defense against potential predators. Perhaps, in a flock comprising several individuals, the members judged themselves capable of performing their own anti-predator defense without any assistance from the Rufous Vanga. In stark contrast to these other member species, the size of conspecific flocks of the Rufous Vanga decreased during the non-breeding period. It would possibly be more appropriate to interpret the decrease in the flock size of Rufous Vangas within our research area as part of an inter-annual change, rather than as any regular seasonal pattern (Fig 3-7).

In general, birds schedule their breeding to coincide with the peak abundance of food materials on which to raise their chicks. For birds in temperate forests, caterpillars are the most important food resource and are most abundant in the canopy during May-June, which is the chick-raising period[12]. Likewise, according to Dr Mizuta, who was studying the breeding ecology of the Madagascar Paradise Flycatcher, the peak abundance of flying insects at our study area occurs during November-December, which is also the breeding period[18]. As such, the food resources are less abundant during the non-breeding period, and thus, the principal purpose for birds joining a mixed-species flock would alternate between taking advantage of anti-predator defense and efficient foraging. In reality, it was only during the non-breeding season that we observed birds fighting over a single insect food item. Considering the above explanation to be valid, the presence of the Rufous Vanga in the mixed-species flock during the non-breeding season is not as useful to the other members during this period because the ground surface where the Rufous Vanga primarily forages is not the optimal foraging site for other bird species, which forage primarily in the canopy.

As noted above, the efficiency of members increased when foraging in mixed-species flocks during the breeding period as well, and defense against predators would remain essential in the mixed-species flock during the non-breeding period. Perhaps the key factor that dictates the different roles of the Rufous Vanga within the theater of the mixed-species flock is whether the priority of the other cast members is anti-predator defense or foraging efficiency. Consequently, during the breeding period when food material is abundant and birds can forage individually or in pairs, defense against predators is the main concern. Therefore, the mixed-species flock is formed by following the breeding group of Rufous Vangas that are extremely vigilant and warn the other members against predators. Conversely, during the non-breeding season, when food resources are less, efficient foraging becomes the primary concern, and birds that utilize similar foraging sites aggregate to form a flock. However, since the Rufous Vanga is less vigilant against predators during this period than during the breeding period, the flock members do not choose this species as a foraging partner. The Rufous Vanga always goes its own way irrespective of whether or not it is the breeding period, and it appears to be totally indifferent to the behavioral changes in the species around them.

The diverse world created by the mixed-species flock

When I was invited to join the research project in Madagascar, Dr Yamagishi, the team leader, allocated me the task of investigating "The effect of interspecific sociological relationships within mixed-species flocks on the society of the Rufous Vanga." Unfortunately, I was unable to provide a clear answer to this difficult question on the basis of my findings during two field research trips. However, if I were to attempt a partial response to this question,

I would state that “mixed-species flocks do not affect the society of the Rufous Vanga in any way.” Moreover, as I have described in this section, the opposite effect was evident. Therefore, I would also state that “the intraspecific society of the Rufous Vanga greatly influenced the interspecific society within mixed-species flocks.”

What are the effects of this interspecific society on the diversity of the bird community in the forest of Jardin A? Twenty-nine species of birds were breeding in the forest. This number is almost equal to that of the birds breeding in broadleaf deciduous forests in the temperate zone, wherein the structure of the strata in the forest canopy is well diversified[14]. However, the forest of Jardin A is fairly poor in terms of structure. In Japanese forests with an equivalent poor structure, perhaps only 15 species or so would be present. The diversity of birds found in the forest of Jardin A is thus comparatively high. When considering community diversity, the evenness of species as well as the number of species composition is important. Let us look at Fig 4-1 once again. This figure lists species’ names in order of abundance of pairs within the community during the breeding period. In temperate bird communities, the line usually forms an concave. As the composition of species grows more even, which in turn signifies a diverse community structure, the dent grows gentler. Birds inhabiting the forest of Jardin A are an excellent example of this trend, and as the curve is a nearly straight line, it implies a high species diversity. Moreover, the regularly flocking species discussed in this chapter are foremost in terms of density composition, and they characterize the community structure. Although highly ranked species in abundance, such as the Souimanga Sunbird *Nectarinia souimanga*, the Common Jery *Neomixis tenella*, the Madagascar Bulbul *Hypsipetes madagascariensis*, and the Crested Coua, participated in mixed-species flocks as supporting actors, their participation rate was not as high as that of the leading actors.

It has been documented that in the tropical forests of South America, species that are members of mixed-species flocks maintain stable communal territories during the breeding season[4,15]. In this case, the number of pairs that join the mixed-species flocks grows equal, and the evenness of species composition increases greatly. While the case of the birds of the Jardin A forest is not that extreme, the more they would try to forage after forming a mixed-species flock during the breeding season, the lesser should the number of available territories. Therefore, a situation wherein a few species dominate numerically does not occur, as is observed in the bird community of the temperate zone wherein mixed-species flocks are not formed during the breeding season. As noted earlier, it is considered that species in tropical forests form mixed-species flocks during the breeding season in order to defend themselves against a high density of predators. Predator-avoidance effects would increase by moving together throughout the year with individuals of different species that have overlapping home ranges. When reliable species such as the Rufous Vanga are part of the group, the self-defense effect becomes invincible. To maintain the diverse bird community in the forest of Jardin A, the presence of the Rufous Vanga is indispensable.

In the mixed-species flocks in the forests of Hokkaido, it was often observed that large-sized birds attacked smaller birds and stole their food. We witness a situation wherein individuals attempt to survive by victimizing other birds. In the mixed-species flock in the Jardin A forest as well, certain selfish behavior was observed whereby the Rufous Vanga, which did not benefit in any manner, was positively followed by other birds while deriving anti-predator benefit. However, although such selfishness is evident, on the whole, a gentle and accommodating society appears to have been established, whereby no bird suffers any costs or demerits. These differences between the mixed-species flocks of the temperate-zone forests and those inhabiting the forest of Jardin A can be likened to the differences existing between the hectic Japanese society and the relaxed society of the people of Madagascar.

References

1 Hino, T. (1993) Interindividnal differences in behavior and organization of avian mixed-species flocks. *Mutualism and Community Organization* (eds. H. Kawanabe, J. E. Cohen, & K. Iwasaki), pp. 87-94, Oxford University Press, Oxford.

2 Hino, T. (2002) Breeding bird community and mixed-species flocking in a deciduous broad-leaved forest in western Madagascar. *Ornithological Science*, 1: 111-116.

3 Bell, H. L. (1983) A bird community of lowland rainforest in New Guinea: Mixed-species feeding flocks. *Emu*, 82: 256-275.

4 Jullien, M. & Thiollay J. M. (1998) Multi-species territoriality and dynamics of neotropical forest understorey bird flocks. *Journal of Animal Ecology*, 67: 227-252.

5 Hino, T. (1998) Mutualistic and commensal organization of avian mixed-species foraging flocks in a forest of western Madagascar. *Journal of Avian Biology*, 29: 17-24.

6 Hino, T. (2000) Intraspecific differences in benefits from feeding in mixed-species flocks. *Journal of Avian Biology*, 31: 441-446.

7 Krebs, J. R. (1973) Social learning and the significance of mixed-species flocks of chickadees (*Parus* spp.). *Canadian Journal of Zoology*, 51: 1275-1278.

8 Caraco, T. (1979) Time budgeting and group size: a theory. *Ecology*, 60: 611-617.

9 Brockmann, H. J. & Barnard, C. J. (1979) Kleptoparasitism in birds. *Animal Behaviour*, 27: 487-514.

10 Swynnerton, C. F. M. (1915) Mixed bird parties. *Ibis*, 67: 346-354.

11 Lack, D. (1954) *The Natural Regulation of Animal Numbers*. Clarendon Press, Oxford.

12 Murakami, M. (2002) Foraging mode shifts of four insectivorous birds under temporally varying resource distribution in a Japanese deciduous forest. *Ornithological Science*, 1: 63-

69.

13 Mizuta, T. (2002) Seasonal changes in egg mass and timing of laying in the Madagascar Paradise Flycatcher *Terpsiphone mutata*. *Ostrich*, 73: 5-10.

14 Hino, T. (1990) Palaearctic deciduous forests and their bird communities: comparison between East Asia and West-central Europe. *Biogeography and Ecology of Forest Bird Communities* (ed. A. Keast), pp. 87-94, SPB Academic Publishing, Hague.

15 Munn, C. A. & Terborgh, J. W. (1979) Multi-species territoriality in neotropical foraging flocks. *Condor*, 81: 338-347.

Social organization of the Rufous Vanga

In the society of this bird, a pair is accompanied by son(s). Please note the color rings.

Plate II-1 Nests of Rufous Vanga are built in the first fork of trees. To prevent access by snakes, the tree trunk was wrapped with plastic sheet, and petroleum jelly was applied on the sheet. A nest is visible above the tree trunk with plastic sheet.

Plate II-2 Brown lemur, one of the natural enemies of the Rufous Vanga.

Plate II-3 The most common clutch size is four.

Plate II-4 An extreme case of asynchronous hatching. Note the difference in sizes of chicks. Some eggs are remain, waiting to hatch.

Plate II-5 Rufous Vanga, ready to fledge.

a

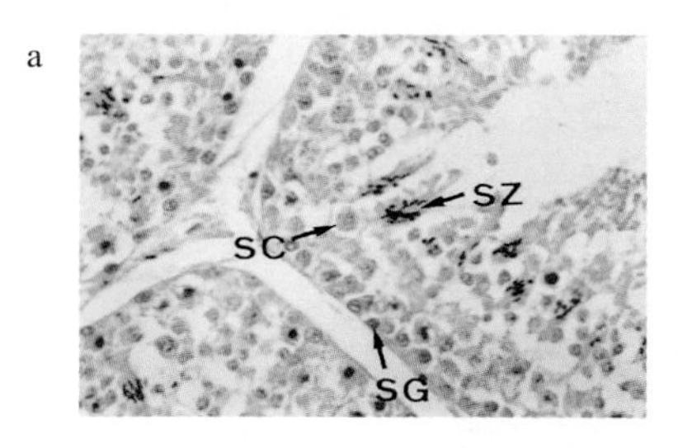

Plate II-6 Development of male testis in a group
Micrographs of testies of an adult male (a), spotted throat male 1 (b), and spotted throat male 2 (c).
SG: spermatogonium
SC: spermatocyte
SZ: spermatozoa
(from Yamagishi et al. 2002)

b

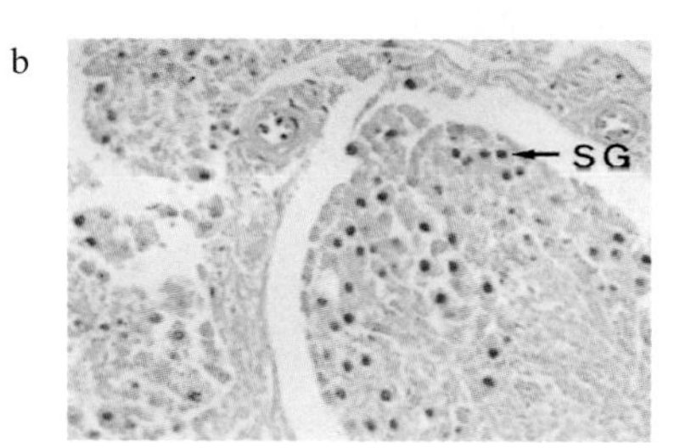

c

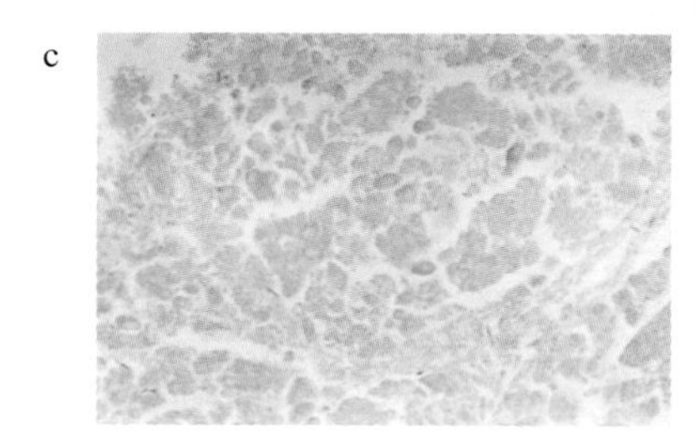

Plate II-7 A spotted throat male helper eventually becomes an adult male with an intensive black throat. Now he looks almost identical to his father (left).

The breeding biology of the Rufous Vanga

Kazuhiro EGUCHI

Breeding season

The Rufous Vanga can be found in areas ranging from the wet tropical rainforests in eastern Madagascar to the deciduous dry forests in the western region (Figs. 9-1 and 9-2). In the western region, the breeding season extends from September to January. Nest building begins during late September. Around this period, breeding pairs can be seen moving from tree to tree or perched in the fork of branches, searching for a suitable nest site. However, the sex that selects the nesting place and the factors that influence the selection of a nest site remain unknown. In most cases, nests are built in the first fork above the ground (Plate II-1). Egg laying begins in October (Table 5-1).

If nesting fails, pairs resume building a replacement nest within a week, repeating their breeding attempts up to four times. Of the 207 breeding groups observed from 1994-1999, 95 groups built only one nest in a season, 87 built nests twice, 19 did so three times, and six did so four times. Thus, less than half the groups succeeded in breeding at their first attempt. More than half of the clutches laid in November, and all the clutches that laid in December were replacement clutches (Table 5-1). Egg laying ends in December; we have not yet observed egg laying in or after January. This species produces only one brood annually.

In the western region of Madagascar, the rainy season begins during the second half of November, and it rains daily in December (Fig 2-5). In regions characterized by a dry climate, where the alternation of the rainy and dry seasons is distinct, the rainfall at the beginning of the season acts as a proximate factor for breeding in the case of certain bird species. For example, the Silvereye *Zosterops lateralis* in northern Australia begins breeding synchronously

Table 5-1. Monthly distribution of laying dates of the Rufous Vanga (modified from Eguchi et al. 2001).

Year	October	November	December	Study period
1994				Oct.-Dec.
First nests	5	0	0	
Replacement nests	2	6	0	
Total	7	6	0	
1995				Aug.-Jan.
First nests	14	9	0	
Replacement nests	3	8	4	
Total	17	17	4	
1996				Sept.-Jan.
First nests	19	4	1	
Replacement nests	0	8	5	
Total	19	12	6	
1997				Sept.-Jan.
First nests	20	9	0	
Replacement nests	2	19	3	
Total	22	28	3	

immediately after the first rainfall at the beginning of the rainy season[2]. The Zebra Finch *Poephila guttata* living in arid areas begins breeding two to four months after the heavy rainfall at the beginning of the breeding season. The rain promotes the sprouting and growth of plants and, in turn, an increase in the number of invertebrates, the primary food resource for these birds. The Zebra Finch appears to time the onset of breeding in such a manner that the period of chick rearing coincides with the period when food resources are abundant[3].

In western Madagascar, the Rufous Vanga begins breeding approximately two months before the first heavy shower of rain; therefore, for this species, rainfall is not a proximate factor for the onset of breeding. The Rufous Vanga feeds upon scorpions and centipedes in the leaf litter as well as on herbivorous insects (e.g., caterpillars) in the foliage. Since the abundance of these terrestrial invertebrates is not influenced by rainfall, it appears that unlike other insectivorous birds, for the Rufous Vanga, the onset of breeding is not related to rainfall. On the contrary, an extended period of rainfall makes it difficult for parent birds to forage and increases the risk of chick mortality through hypothermia. This could explain why egg laying is completed before January. Thus, since egg laying is limited to the period between October and December, the breeding season of this species is relatively short[4].

Chicks of the Rufous Vanga fledge well before their primary feathers have developed completely and they are fully capable of flying (Plate II-5, Fig 5-1). In species that live under conditions of high nest-predation, it is advantageous if chicks fledge as soon as they are able to move out of the nest. However, it takes

Fig. 5-1 A fledgling banded with color rings.

longer to rear chicks after fledging. Although data available on the duration for which fledglings are dependent on their parents is limited, we observed that a chick fledging in Group-H on November 5, 1996 was accompanied by its parents even 20 days later. That year, Group-H was the first to begin breeding. This indicates that even among the groups that began breeding early, parental care for fledglings continued until at least the end of November. If a one-month fledgling dependence upon the parents is typical, then it appears that a second breeding during the period when environmental conditions remain favorable is impossible.

Nest shape and nest site

Nests are bowl-shaped and built in a tree fork approximately 4 m above the ground (Fig 5-2). Constructed from assorted materials, such as moss, twigs, dead leaves, tree chips, tree bark, and plant fibers, these nests are bound together with spider webs (Table 5-2). Both sexes participate in nest building. During the initial stages of nest building, birds carry spider webs on their neck and throat and daub them in the fork of the tree. After several spider webs have accumulated, they pile nest materials atop this layer and use fine twigs as a nest lining. When sufficient materials have been piled up, they shape the nest bowl

Fig. 5-2 The bowl-shaped nest of the Rufous Vanga built in a tree fork. An adult male is incubating.

Table 5-2. Relative frequency (%) with which different nest materials were carried to the nest as a function of the stage of nest building. (modified from Eguchi et al. 2001).

Stage	Nest materials							None	Total
	Spider web	Twig	Bark	Chips	Moss	Leaf	Fiber		
Early	64.9	8.8	4.4	11.9	1.3	1.6	2.5	4.7	319
Middle	59.4	11.1	4.8	12.8	1.7	0.4	2.4	7.2	414
Late	58.0	19.7	3.0	3.0	0.4	1.5	3.0	11.4	264

by pressing the wall and floor of the nest with their breast. Although a large amount of materials is used, nest construction is not so elaborate, and the end product resembles a simple pile of materials. A similar nest is built by the Hook-billed Vanga *Vanga curvirostris* and the Helmet Vanga *Euryceros prevostii* (Chapter 8).

In 1997, we observed an instance of particularly unusual nest-building behavior in Group-L, which included an adult helper (helpers are usually males, unless otherwise noted). Both the male and the female of the pair participated actively in nest building. However, they frequently carried nest materials to two different places on the same tree, approximately 50 cm vertically apart from each other. Each bird carried nest materials in almost equivalent amounts to these two nesting places. We observed no division of nesting places between the male and the female. Although the helper was less frequently involved in nest building than the breeding pair, it also carried nest materials in equal amounts to the two different places.

I was baffled by this behavior. Before this event, we had sometimes observed that after a sudden interruption in nest building, a pair or a group would resume building a new nest at a different location. We also noted instances wherein helpers carried nest materials to another place after completing the building of a nest; however, such nest-building behavior was arbitrary and soon ceased. In contrast, in Group-L, both nests were built to a stage of near completion after a period of several days. Eventually, however, only the lower nest was used for breeding, and the upper nest was deserted during the stage of interior construction. Although the nest itself is not sophisticated, nest building is not a trivial or easy task. Moreover, this instance did not involve an adult helper who had attained sexual maturity and was building the nest for its own breeding activity. Given this situation, it would appear fruitless to build a nest that would not be used in future. Could it be that each parent held a different opinion regarding the precise position of the nest and that they did not reach a compromise?

Egg laying and incubation

A female lays an egg daily until the clutch is complete. A clutch typically consists of four eggs (Plate II-3); however, in the replacement clutches, clutch size is marginally smaller (Table 5-3)[1]. Incubation begins on the day that the first egg is laid. Climbing up the tree daily to observe the nest in order to identify the day when the eggs will hatch is very tiring for the researcher. Moreover, this disturbs the bird's normal breeding activity. To avoid such problems, we observed the contents of the nest from the ground using a mirror attached to a telescopic fishing rod (Fig 5-3). On an average, the first chick of a brood hatched 16. 2 days after the last egg was laid (14-19 days: data from 62 nests, 1994-1997), and the mode was 16 days. Since incubation begins on the day that the first egg was laid, the duration of the incubation period is calculated as 16 days plus the number of eggs minus one: the mean clutch size is four, and thus the mean duration of the incubation period is 19 days.

In the period immediately after the beginning of the egg-laying period, pair males contributed the most to incubation activities. However, after the end of egg laying, both pair members contributed almost equally (Fig 5-4). In the case of groups with adult helpers, during the period when the first and second eggs

Table 5-3. Mean ± SD (n) of clutch size, number of hatchlings and number of fledglings of the Rufous Vanga (modified from Eguchi et al. 2001).

	First nests[1]	Replacement nests[1]
Clutch size	3.7 ± 0.62 (n=68)	3.5 ± 0.66 (n=57)
No. of hatchlings	3.0 ± 0.89 (n=52)	3.1 ± 0.62 (n=27)
No. of fledglings	2.4 ± 1.17 (n=31)	2.0 ± 1.18 (n=11)

[1] Data are combined for all years (1994-1999).

Fig. 5-3 Nest checking with a mirror attached to a fishing rod.

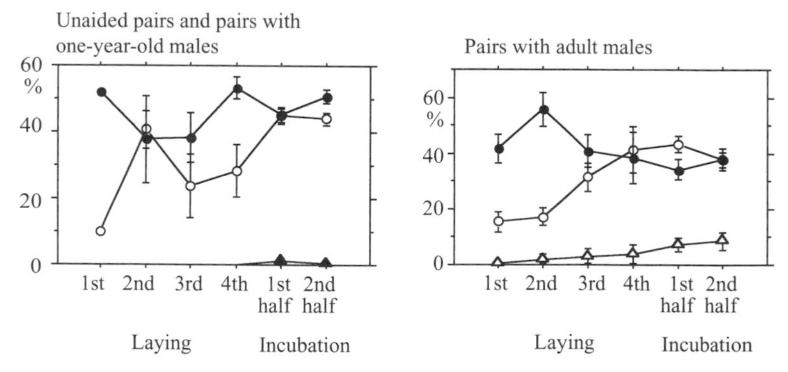

Fig. 5-4 Changes in the amount of time (%) that individual Rufous Vangas sat on their nests during nest watches. Open circles = breeding females. Closed circles = dominant males. Open triangles = adult auxiliary males. Closed triangles = one-year-old males. Data for one-year-old males in groups with adult auxiliary males are not shown because they did not visit nests. Vertical bars indicate SE. (from Eguchi & Yamagishi (2002)[4])

were laid, pair males usually spent 40-50% of the observation time engaged in incubation, whereas pair females spent less than 20% of the time in these activities. During the second half of the laying period, however, the time spent in incubation by pair females increased, and after the termination of egg laying, the mean percentages of time spent in incubation were almost equivalent between pair members (Fig 5-4).

Such incubation behavior by males is puzzling. Mate-guarding behavior by pair males during the period when females are fertilizable has been observed in several species. For example, in a polyandrous group of the Dunnock *Prunella modularis*, the dominant male associates closely with its mate and prevents subordinate males from copulating with the female during the egg-laying period when she is receptive[5]. In the case of the Rufous Vanga, we have evidence suggesting that an adult helper copulated with the pair female and sired chicks[6]. In order to prevent extra-pair copulations during the laying period, it is important for the pair males to associate closely with the pair females and guard against adult helpers. Indeed, pair males often chased away adult helpers whenever they found the helpers near the breeding females (Fig 5-5 a, b). However, instead of associating closely with the pair females, the pair males often sat on the nest after the first egg laying. On the other hand, the pair females were often away from their nests. On those occasions, the females were approached by the helpers and were fed by them. In most cases, females did not refuse to receive food; in fact, they occasionally snatched it from the helpers.

In Group-TX in 1997, in which fertilization by the adult helper was suspected, the pair male and female sat on the nest for 57% and 18% of the observation time on the day the first egg was laid; 79% and 7% on the day the second egg was laid; 38% and 48% on the day the third egg was laid; and 10% and 64% on the day the fourth egg was laid, respectively. During this period, while the helper male seldom sat on the nest, the pair male continued sitting on it and frequently chased away the helper male whenever he was seen in close proximity of the female. However, the pair male continued to incubate and did not associate closely with the female. Viewed from the conventional perspective of mate guarding to prevent extra-pair copulation, the behavior of this pair male is puzzling. Most of the helpers of the Rufous Vanga are the sons of pair males and, less often, their brothers (Chapters 3 and 6). On the basis of the theory of inclusive fitness, copulation by helper males that are closely related to the pair males may be less detrimental for the latter than copulation by helpers that are unrelated to them. In 1995, an adult male invaded the territory of Group-T, which comprised an unaided pair, and stayed there until the termination of breeding. However, the pair male in Group-T attacked the invader male more violently as compared to similar attacks in other groups. Although the

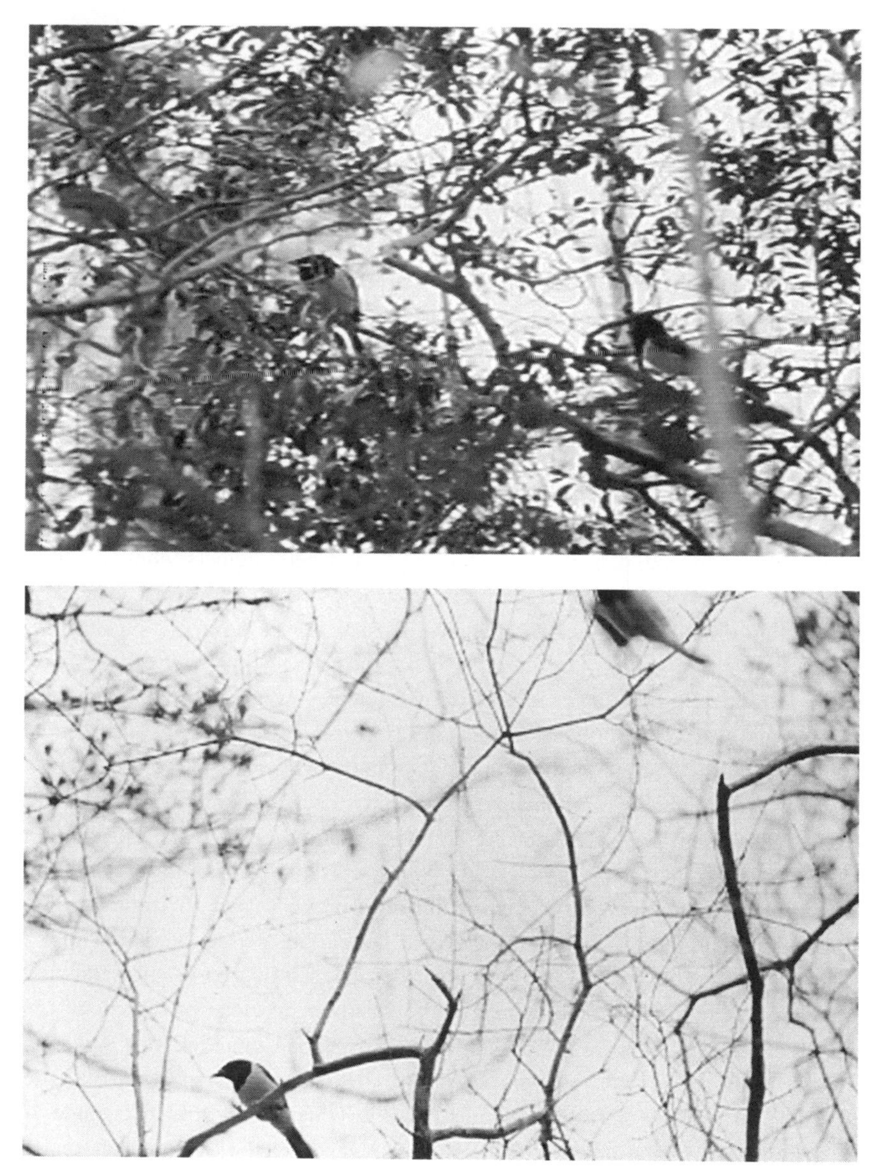

Fig. 5-5 The father becomes aggressive to his son when the two-year-old son becomes black throated and a new female has joined them.
a: female (left), father (center), and son (right) are in a tense situation.
b: the father (right) is attacking at his son (left).

genealogies are unknown in this case, the proximity of relation between the pair male and helper may be one factor that influences the aggressiveness and incubation behavior of the pair males.

The behavior of this invader male is also puzzling. In this group, egg laying was completed on October 29th. The invader male was not banded. However, no unbanded male had been observed in the vicinity of the territory of Group-T until October 30th. On November 2nd, this male was violently attacked by the pair male in the territory. However, on November 4th, the invader male sat on the nest of Group-T and incubated the eggs. The duration of incubation was long, 20% of observation time. On this day, the pair male was not observed. This suggested that the invader male had usurped the territory. On November 7th, however, the pair male reappeared, resumed incubation, and violently attacked the invader male. Ultimately, the pair succeeded in fledging chicks. However, the invader male stayed at the periphery of the territory. After the brood had fledged, it resumed its pursuit of the female, and it was attacked by the pair male whenever it approached the female. Thus, it is likely that the invader male incubated a clutch that was not its own. The behavior of the female is also puzzling: it behaved normally and did not exhibit any hostile behavior toward the invader male on the day that it incubated the eggs.

Pattern of hatching

The embryos of birds begin to develop when eggs are warmed over the threshold temperature that is termed "physiological zero." Before the onset of development, eggs can survive low temperatures; however, if egg temperature is lowered after it exceeds the physiological zero, the eggs will perish due to hypothermia. Therefore, after the onset of embryonic development, the pair has to incubate constantly. If the pair begins incubation immediately after the first egg is laid, the eggs will hatch in the order in which they were laid (asynchronous hatching).

Usually, the timing of the onset of incubation determines the hatching pattern. In the case of the Rufous Vanga, since incubation begins when the first egg is laid (Fig 5-4), eggs should hatch asynchronously (Plate II-4). However, hatching in this species is not exactly asynchronous. Various hatching patterns, ranging from complete asynchronous hatching to complete synchronous hatching, are observed (Table 5-4). The fact that males initiate incubation immediately after the first egg is laid suggests that males induce asynchronous hatching. Few studies have examined the sharing of incubation in birds, and no study has investigated which sex initiates incubation[7]. In songbirds, the situation wherein

Table 5-4. Hatching pattern (Eguchi & Yamagishi 2002)[4].

Nest	Breeding[1]	No. of hatchlings				BS[2]	CS[3]	Laying date	MIP[4]
		Day 1	Day 2	Day 3	Day 4				
1995									
H	py	1	2			3	3	10/9	16
D	pm	2	?	?	?	4	4	10/13	16
Y	py	2				2	3	10/22	17
C	pm	1	1	1		3	4	10/22	16
CX	py	2	1			3	3	10/24	18
EX	p	2	1			3	4	10/25	17
E	p	1	2			3	4	10/26	16
T	p	1	1	1	1	4	4	10/26	17
F2	p	2	1			3	4	10/28	16
G	p	1	1	1		3	4	10/28	16
M2	p	1	2	1		4	4	10/29	16
CC	p	1	1			2	3	11/1	17
L	py	2	1	1		4	4	11/7	16
EE	p	2	1			3	3	11/9	17
IX	p	3				3	4	11/23	17
1996									
IX	p	1	1			2	4	10/10	16
L	pm	3				3	4	10/24	16
T	p	4				4	4	10/25	17
M	p	2				2	4	11/12	18
1997									
CX	p	1	2	1		4	4	10/22	16
IX	p	3				3	4	10/27	14
JX2	py	2	1			3	4	11/4	16
RX4	py	1	2			3	3	12/10	16
T2	p	3				3	4	?	?

1: p = pairs without helpers, pm=pairs with adult helpers, py = pairs with one-year-old helpers.
2: BS = brood size.
3: CS = clutch size.
4: MIP (Minimum Incubation Period) = interval from clutch completion to first egg hatching.

incubation begins from the time the first egg is laid and wherein the male incubates to a greater extent than the female is extremely rare[8]. Initiation of incubation by the male may be a feature that is unique to the Rufous Vanga.

Several hypotheses have been proposed regarding the adaptive significance of the early onset of incubation and asynchronous hatching[7]. One hypothesis states that since chicks fledge in the order of hatching, the risk of the entire brood suffering nest predation decreases, and the possibility that at least one chick can fledge increases[9,10]. In the case of synchronous hatching, since the onset of incubation is delayed until clutch completion and eggs laid earlier are in the nest for a longer time than the minimum period of incubation, there is an increased risk that these eggs will suffer mortality through predation and adverse weather conditions. If the parents sit on the nest from the onset of egg laying, they can protect the eggs from adverse weather and predators. Furthermore, since the period between laying and fledging is at a minimum for each egg, and since chicks fledge in sequence, the risk of the entire clutch or brood being predated will be lowered.

However, in the case of asynchronous hatching, although the period for which the eggs are in the nest decreases, the period for which chicks occupy the nest increases. Hungry chicks may noisily beg for food and thereby attract the attention of predators; thus, asynchronous hatching can increase the predation rate on chicks and consequently lower breeding success[7]. In the case of the Rufous Vanga, predation on eggs and chicks is extremely high. On the basis of data from a six-year study, while 133 (49%) of 272 nests in which eggs were laid failed before hatching, 61 (48%) of 127 nests failed before fledging, almost equivalent rates of failure. Since the incubation period is slightly longer than the nestling period, daily risk of predation on eggs may be higher than that on chicks. However, this difference may be small. When incubation begins after clutch completion, the first egg remains in the nest 3-4 days longer than the last egg, and the risk of predation increases. In an environment wherein predation on eggs or chicks is high, it is advantageous for the parents to begin incubation immediately after the first egg is laid and to induce asynchronous hatching. However, in the case of the Rufous Vanga, early onset of incubation does not always induce asynchronous hatching. Furthermore, in more than half of the fledging broods that we observed, all the chicks of the brood fledged synchronously. Hence, early onset of incubation does not induce asynchronous fledging and probably does not decrease the risk of predation.

This being so, why is the onset of incubation early in the case of the Rufous Vanga? Among birds, unincubated eggs generally exhibit a decreased hatchability[7]. In western Madagascar, the maximum temperature exceeds 35°C (Fig 2-5) and the sunlight is strong. If the physiological zero is around 28°C as reported[7], then environmental conditions alone might trigger embryonic development in some clutches even without incubation. In such clutches, if the egg temperature is not maintained above the physiological zero, the embryos could die. In this area, the ambient temperature decreases to below 20°C due to radiative cooling at night and dawn (Fig 2-5); conversely, direct sunlight raises the egg temperature above the desired level. Since the effect of high temperatures on mortality is more severe than that of low temperatures, incubation may be also necessary to protect the eggs from direct sunlight. In the case of the Rufous Vanga, protecting the eggs from adverse weather conditions and predation may be the primary purpose of incubation from the time the first egg is laid, and the patterns of hatching and fledging that result from this incubation pattern might not be so important to the parents.

While pair males contribute toward incubation to a greater extent than females at the beginning of incubation, and incubation duties are almost equivalent between the sexes after clutch completion, females do contribute more to brooding than do the males (Figs. 5-4 and 5-6). The greater contribution of the pair males to incubation at the beginning of this process is noteworthy behavior. If the early onset of incubation is necessary to protect the eggs, and if incubation is a cooperative task, then the contribution should be equivalent between sexes. However, if females have to spend more time foraging to gain energy to produce eggs during the egg-laying period, then it is possible that the pair males contribute more to incubation during this period by way of division of labor. However, the greater contribution of females toward brooding is not explicable in terms of the hypothesis of division of labor.

An alternative perspective on the adaptive significance of asynchronous hatching was provided by T. Slagsvold and J. T. Lifjeld[11], who explained this phenomenon in terms of sexual conflict rather than cooperation between sexes. On the basis of their comprehensive study of bird species wherein only the female incubates, they posited that in such species, the females induce asynchronous hatching by beginning incubation early in the egg-laying period. The females behave in this manner in order to extend the period during which the males feed nestlings and brooding females. This provides some benefits to the females, such as the fact that the increased provisioning duties of the male reduces the feeding duties of the female and also makes it difficult for the pair males to seek extra-pair copulation. In the case of the Rufous Vanga, the percent contribution of the pair male and female toward incubation and brooding changed in relation to the breeding stage. By observing this changing pattern, one can understand that although the males initiate incubation, which results in asynchronous hatching, the early onset of incubation does not provide them any benefits during the rearing of chicks.

What benefits do males gain by initiating incubation immediately after the first egg is laid, a behavior that prevents them from guarding their mate? If early incubation does not always induce asynchronous hatching, it is unlikely that males derive any benefit in the post-hatching stages. In a situation wherein brood mortality increases unless either or both members of the pair incubate from an early stage, and wherein the females frequently leave the nest in order to forage, males must incubate in order to retain the existing level of fitness gain. As noted earlier, since helpers are usually the sons of the pair males, the males derive some indirect benefit even if they lose their paternity to the helper.

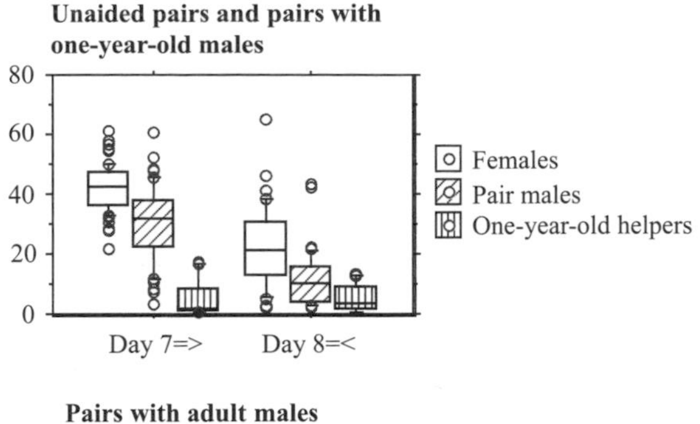

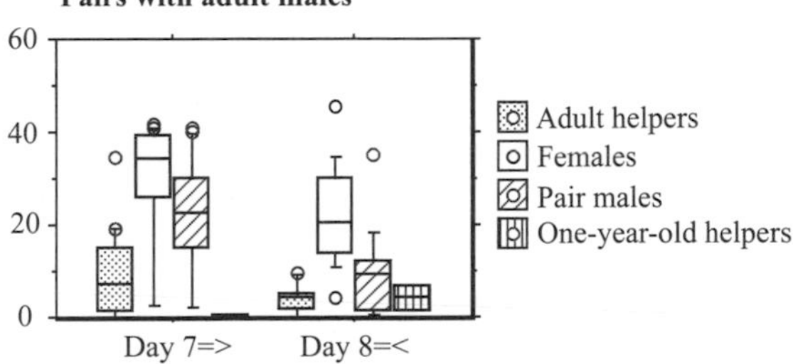

Fig. 5-6 A comparison of percent time spent in nest attendance. Most of the attendance time was spent in brooding during the earlier half of rearing stage and in shading during the later half.

Care of chicks

The mean duration between the hatching of the first chick (day 0) and the first fledging is 14. 8 days (12-17 days: based on data collected from 1994-1997). Chicks fledge before their primary feathers have developed completely and before they are able to fly (Plate II-5, Fig 5-1).

Both pair members participate in the care of chicks and visit the nest at an equal frequency (Fig 5-7)[4]. However, within a week of hatching, the female spends more time brooding than the pair male (Fig 5-6). Even in nests with old nestlings, the female spends more time at the nest than the male. Although the nest-visiting frequency is equal for both pair members, the female often visits without food; hence, the female provisions the chicks at a lower frequency than the male (Fig 5-7). In summary, the male spends more time at the nest during the egg-laying and incubation periods, whereas the female does so during the nestling period. While division of labor between the sexes may occur to a certain extent, no distinct division, such as within and outside the nest, is observed.

Even in a nest that contains old nestlings, parents often positioned themselves

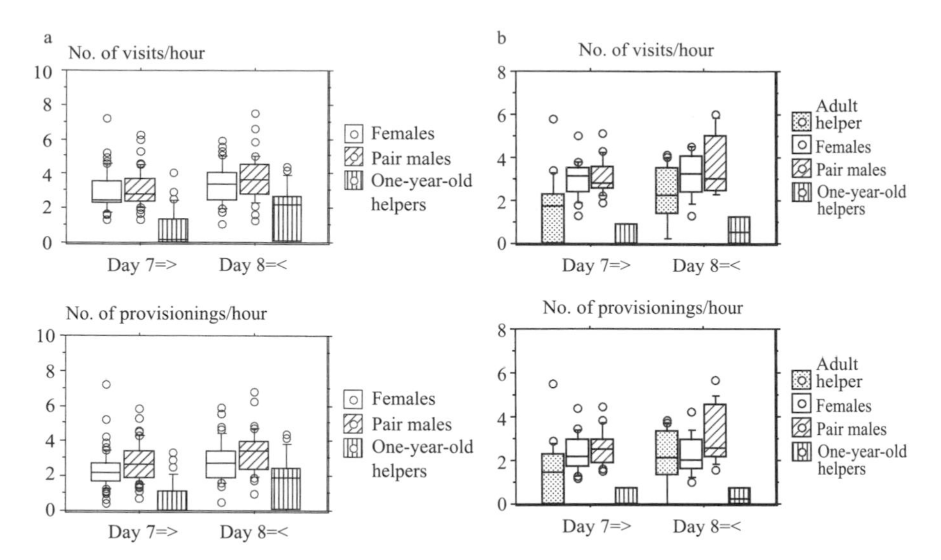

Fig. 5-7 Numbers of visits and provisionings.
a:Pairs with one-year-old males,
b:Pairs with one-year-old males and adult males

on the nest rim to shade the chicks from direct sunlight. However, both sexes spent more than 30% of their time on nest attendance, which suggests that shading was not their exclusive purpose. Prevention of nest predation or interference might be another reason for their stay at the nest: during nest watches, we often observed that individuals sitting on the nest rushed out of the nest and attacked animals approaching it.

Table 5-5 shows the prey items that comprised the diet of the nestlings. Major food items are caterpillars, crickets, cicadas, and spiders; reptiles (chameleons and lizards) and invertebrates (centipedes and scorpions) that were found in the leaf litter were also brought to the nest frequently. The Rufous Vanga often forages by pouncing on prey on the ground and on cicadas on tree trunks[12]. There is no sexual difference either in terms of the size of prey brought to nests or in prey type (Fig 5-8).

In the case of the Rufous Vanga, yearling or adult males join the breeding pairs and assist them as helpers. In groups with helpers, the helpers assist in providing food to the chicks. Their contribution to provisioning is substantial and can account for 30% of all feedings (Fig 5-7b). Helpers also attempt to assist in nest building and incubation, but they are often attacked by the breeding pair; therefore, their realized contribution is low.

A comparison of the feeding frequency between yearling helpers and adult helpers reveals that adult helpers feed young chicks more frequently than do

Table 5-5. Prey items carried to chicks (data collected from 1994 to 1997)

Reptiles		Insects		Arthropods other than insects	
Lizards	31	Caterpillars	220	Spiders	112
Chameleons	28	Moths	25	Centipedes	35
Geckos	10	Crickets	204	Scorpions	24
		Miscellaneous grasshoppers	49	Miscellaneous arthropods	22
		Cicadas	152		
		Miscellaneous insects	103		

yearling helpers (Fig 5-9). This is because when the chicks are still young, the feeding frequency of yearling helpers is low and several yearlings never visit the nest. However, as the chicks grow, the number of individuals that participate in feeding increases, as does the feeding frequency of yearling helpers. The difference between yearlings and adult helpers then disappears (Fig 5-9). The contribution of adult helpers toward provisioning becomes almost equal to that of breeding individuals (Fig 5-7). Thus, helpers work hard when the demand of chicks increases. Interestingly, when they feed young chicks, yearling helpers often carry smaller food items; however, once the chicks are 8 days old or more, the variance in the size of prey brought by different age classes of helpers disappears (Fig 5-8). Females often visit the nest without food and sit on the nest. In comparison, the frequency at which yearling helpers visit the nest with food is the same as those of the breeding males and adult helpers. Moreover, the contribution of young helpers toward provisioning older chicks equals that of these two categories.

We have noted some interesting observations with regard to foraging. The first concerns the kleptoparasitic behavior of other birds. The Rufous Vanga often brings adult moths to the chicks, but it appears that these birds are not adept at catching flying insects. Unless the substrate is sturdy, the Rufous Vanga would be unable to maintain its body position when flying over prey items. Therefore, a large number of the insects that they capture are those found resting on tree trunks and boughs, such as cicadas; however, a few flying insects, such as moths, are also caught. In contrast, the Long-billed Greenbul is skilled at catching flying insects. It often forages near the ground, and its foraging niche overlaps that of the Rufous Vanga[13]. At times, the Rufous Vanga persistently chases the Greenbul as it catches moths and then usurps the food item. This is a clear case of kleptoparasitism. In Chapter 4, we mentioned that such behavior is rare; however, this does not imply that it never occurs.

Another example is the act of following foraging birds and capturing insects that get disturbed by the activities of the foraging individual: exploiting other birds as beaters. In 1997, while I was conducting a census in the forest, I came

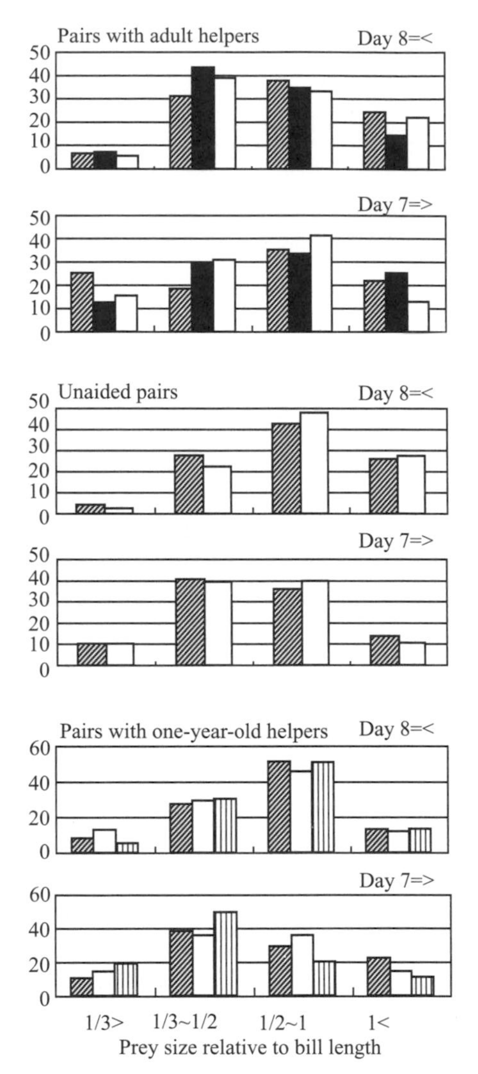

Fig. 5-8 Size distribution of prey items carried to chicks. Hatched columns are for breeder males, closed columns for adult helpers, open columns for females and striped columns for one-year helpers. Ordinates are percentage.

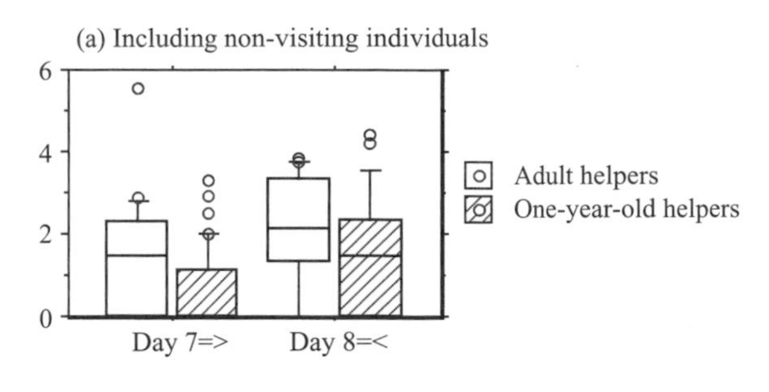

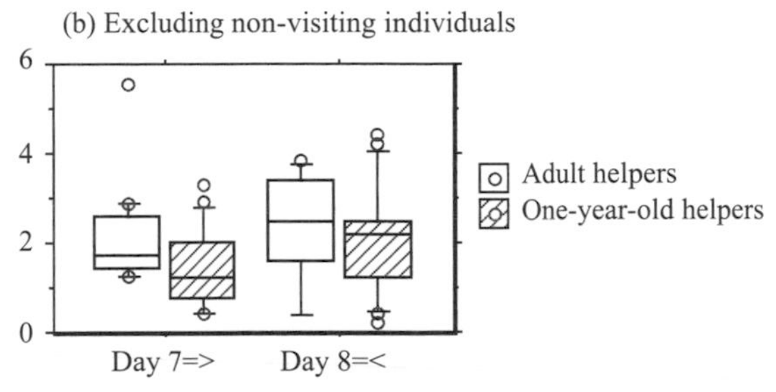

Fig. 5-9 Comparisons of provisioning frequency. Differences are significant only in the provisioning to young chicks in both cases.

across a group of three White-breasted Mesites. This species is endemic to Madagascar and very rare; therefore, I observed them carefully. I then spotted a female Rufous Vanga perched on a branch a few meters away. I saw the Rufous Vanga fly to the ground near the Mesites and catch a prey item. It then returned to the perch and ate the prey but did not leave. Later, it once more flew down close to the Mesites and caught prey. Mesites, like rails, search for food by disturbing leaf litter on the ground. They continued foraging, moving along slowly in the forest. The Rufous Vanga followed as the Mesites moved, and it often flew to the ground to catch prey. The Rufous Vanga usually forages on trees. However, on this occasion, it foraged solely on the ground. The Rufous Vanga was following the Mesites in order to catch mole crickets and other crickets that jumped out from the leaf litter when the Mesites disturbed it. In the one hour and 54 minutes during which I observed this following behavior, the Mesites traversed approximately 350 m and the Rufous Vanga made 14 foraging attempts[14]. An additional feature of this episode was that a Greenbul followed the Mesites and the Rufous Vanga during the entire period. This Greenbul caught flying insects that were disturbed from the undergrowth due to the foraging activities of the Mesites and the Rufous Vanga. Therefore, as documented in Chapter 4, one of the reasons underlying the formation of a mixed-species flock is concerned with the exploitation of the activities of other species.

Breeding success

Breeding success is low, and on an average, only 0. 25-0. 33% of pairs that attempt breeding succeed in fledging their chicks (Table 5-6)[1]. Even if breeding is repeated, scarcely any improvement is observed. In most failed nests, the entire clutch of eggs or chicks disappeared at the same time; therefore, it is believed that they were predated. Although we never directly observed predation, we did observe the Rufous Vanga attacking the snake *Ithycyphus miniatus*, the lizard *Oplurus cuvieri*, the Brown Lemur, etc., near the depredated nests (Plate II-2, Fig 5-10). The severity of these attacks suggested that these animals were potential predators. Further, the Rufous Vanga also attacked several raptor species, even in places that were distant from the nest. Raptors are potential predators as well; in 1999, Dr S. Asai observed that a breeding female was caught by the Madagascar Goshawk *Accipiter henstii*. However, it is not known whether the Goshawk also preyed upon the Rufous Vanga's nest contents.

In contrast, brood reduction occurred at only 12 nests (17%) out of 69 nests surviving until a minimum of 10 days after hatching. Nestling mortality often occurred within one week after hatching; it is considered that this happened because younger chicks were crushed or suffocated by their elder siblings. In the case of the Rufous Vanga, serious food shortage is not observed and starvation is unlikely to be an important factor in chick mortality. Breeding success exhibits large annual fluctuations. Breeding success particularly decreased in 1997 (Table 5-6). The precise reason for this decrease is unclear, but most failures occurred due to predation. Therefore, for some reason, the incidence of predation might have increased.

In several species, the height of the nest above the ground is known to affect breeding success. As the season progresses, the height at which the Rufous Vanga builds its nest decreases insignificantly. The height of replaced nests is

Table 5-6. Nesting success (number of pairs) of the Rufous Vanga (Eguchi et al. 2001).

	1994	1995	1996	1997
First nests				
Successful pairs[1]	3	10	14	6
Failed pairs	10	18	19	37
Total number of pairs studied	13	31	33	43
Replacement nests				
Successful pairs[1]	2	4	3	1
Failed pairs	4	4	5	24
Pairs attempting replacement nesting	9	13	14	26

[1] Pairs fledging at least one chick were considered successful pairs.

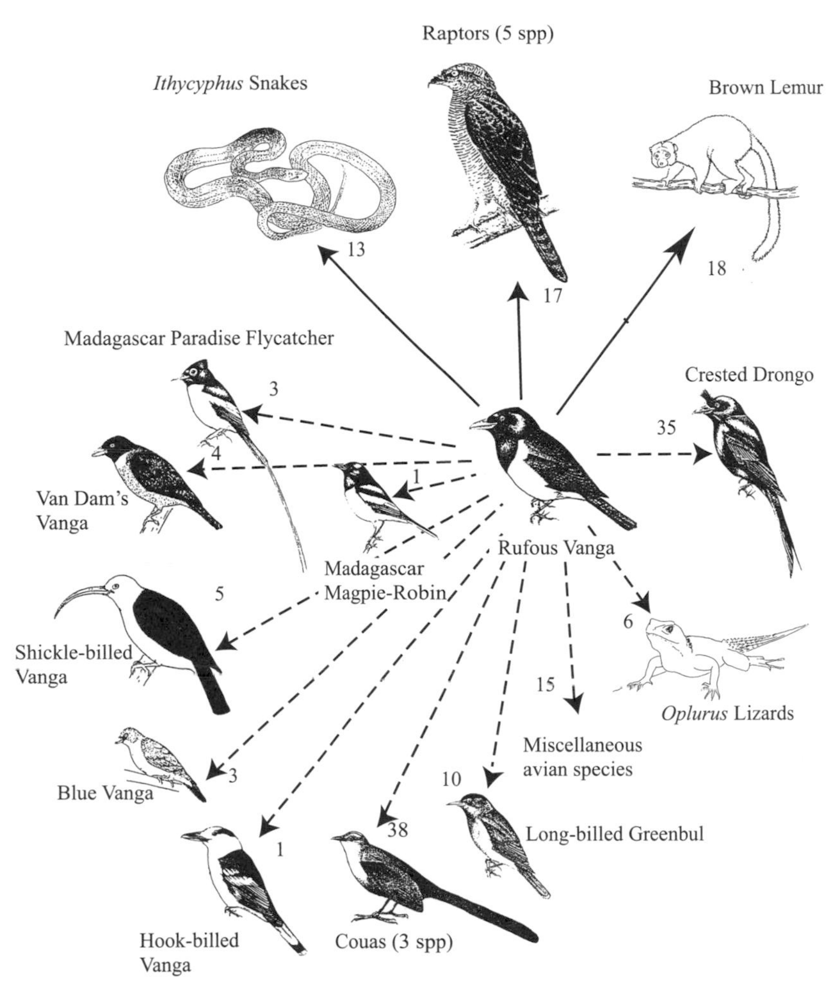

Fig. 5-10 Observations of attacks of the Rufous Vangas against predators and intruders. Solid lines indicate dangerous animals, and broken lines indicate less dangerous animals. Figures are the number of observations during the study from 1994 to 2000 (Asai & Eguchi unpublished data).

also lower than that of the first nests (Table 5-7)[1]. Predation is the primary cause for the breeding failure of the Rufous Vanga, and the change of nest location after failure in breeding could be an adaptive response to reduce predation. However, our data does not support such a conclusion. A comparison of the heights of all first-built nests revealed that the successful nests were 4.0 $\pm$ 1.2 m (n=32) above the ground and the failed nests were 4.3 $\pm$ 1.3 m (n=81) above the ground. Among the replaced nests, the average height of successful nests was 3.

Table 5-7. Seasonal changes in nest height and DBH [Mean ± SD (n)] (modified from Eguchi et al. 2001).

	Sept./Oct.[1]	November	December
Nest height			
First nests	4.3 ± 1.29 (102)	3.8 ± 0.80 (11)	3.5 ± 0.71 (2)
Replacement nests	3.8 ± 1.57 (20)	3.8 ± 1.49 (44)	3.7 ± 1.36 (13)
DBH			
First nests	15.0 ± 5.88 (102)	14.6 ± 6.05 (11)	23.3 ± 18.03 (2)
Replacement nests	12.9 ± 4.29 (20)	13.3 ± 5.09 (44)	15.2 ± 5.79 (13)

[1] Data were combined for September and October because the exact date when nest building started was not determined for some nests.

6 ± 1.5 m (n=10) and that of failed nests was 3.6 ± 1.4 m (n=57) above the ground. Thus, there is no significant difference between them. In the case of the Rufous Vanga, no relationship exists between breeding success and the height of the nest site. Therefore, building a replaced nest at a lower height is not an adaptive response aimed at improving breeding success.

Conclusion

Among populations of Rufous Vangas in the western regions of Madagascar, breeding begins in September, just before the onset of the rainy season, and continues until January, which is the middle of the rainy season. Although these Rufous Vangas inhabit an arid area, breeding is not triggered by the onset of rainfall. One breeding cycle occurs annually. However, should a breeding attempt fail, the pair usually makes another attempt.

Incubation begins after the first egg is laid, but the hatching pattern varies from complete asynchrony to complete synchrony. The major contribution of the male toward incubation from the time the first egg is laid and the absence of any distinct mate-guarding behavior during the egg-laying period are features characteristic to the Rufous Vanga. The contribution of the female toward incubation increases gradually until egg laying is completed; thereafter, the respective contributions of the pair members become almost equal. After the eggs hatch, the female undertakes most of the brooding duties. Even after the chicks grow, the amount of attendance time of the female at the nest is considerable. The onset of incubation activity by the male from the laying of the first egg does not necessarily induce asynchronous hatching; this behavior is therefore not an adaptation toward predation, but it is considered to serve as a measure that prevents hatching failure due to the overheating or cooling of eggs.

The basic breeding unit is a pair. However, at times, yearling males and males aged two years and over also accompany the breeding pair and assist it in breeding. These helpers contribute substantially toward provisioning the chicks, and their assistance can account for up to 30% of the total provisioning activities toward chicks in a given nest.

Breeding success is low, and more than half the pairs engage in rebreeding due to a failed initial breeding attempt. However, even in such instances, the overall breeding success within the population does not improve. A majority of the breeding failures can be ascribed to predation of eggs or chicks. High predation pressure is a feature that is common to other birds in tropical regions. Further research on this particular Rufous Vanga population will help elucidate the relations between the evolution of life history and predation pressure.

References

1 Eguchi, K., Nagata, H., Asai, S., & Yamagishi, S. (2001) Nesting habits of the Rufous Vanga in Madagascar. *Ostrich*, 72: 201-218.

2 Kikkawa, J. & Wilson, J. M. (1983) Breeding and dominance among the Heron Island Silvereyes *Zosterops lateralis chlorocephala. Emu*, 83: 181-198.

3 Hahn, T. P., Boswell, T., Wingfield, J. C., & Ball, G. F. (1997) Temporal flexibility in avian reproduction: patterns and mechanisms. *Current Ornithology* vol. 14, (eds. V. Nolan Jr., E. D. Ketterson, & C. F. Thompson), pp. 39-80, Plenum Press, New York.

4 Eguchi, K. & Yamagishi, S. (2002) Onset of incubation and hatching pattern in the Rufous Vanga. *Journal of Yamashina Institute for Ornithology*, 34: 16-29.

5 Davies, N. B. (1992) *Dunnock Behaviour and Social Evolution*. Oxford University Press, Oxford.

6 Yamagishi, S., Asai, S., Eguchi, K., & Wada, M. (2002) Spotted-throat individuals of Rufous Vanga *Schetba rufa* are yearling males and presumably sterile. *Ornithological Science*, 1: 95-100.

7 Stoleson, S. H. & Beissinger, S. R. (1995) Hatching asynchrony and the onset of incubation in birds, revisited: When is the critical period? *Current Ornithology* vol. 12, (ed. D. M. Power), pp. 115-139, Plenum Press, New York.

8 Haneda, K. (1986) *Life History of Birds*. Tsukiji Shokan Press, Tokyo (in Japanese)

9 Magrath, R. D. (1990) Hatching asynchrony in altricial birds. *Biological Reviews*, 65: 587-622.

10 Clark, A. B. & Wilson, D. S. (1981) Avian breeding adaptations: hatching asynchrony, brood reduction, and nest failure. *Quarterly Reviews of Biology*, 56: 253-277.

11 Slagsvold, T. & Lifjeld, J. T. (1989) Hatching asynchrony in birds: the hypothesis of sexual conflict over parental investment. *American Naturalist*, 134: 239-253.

12 Yamagishi, S. & Eguchi, K. (1996) Comparative foraging ecology of Madagascar vangids (Vangidae). *Ibis*, 138: 283-290.

13 Eguchi, K., Yamagishi, S., & Randrianasolo, V. (1993) The composition and foraging behaviour of mixed-species flocks of forest-living birds in Madagascar. *Ibis*, 135: 91-96.

14 Eguchi, K. (1998) The White-breasted Mesite: Rufous Vanga's beater. *Newsletter of working group on birds in the Madagascar region*, 8: 5-6.

The role of helpers in the Rufous Vanga society

Kazuhiro EGUCHI

Necessary background: why do birds breed cooperatively?

An individual that does not breed but helps a breeding individual is termed a helper, and such a breeding pattern is termed "cooperative breeding." More than 220 avian species (approximately 3%) around the world are known to be cooperative breeders[1]. With regard to several species, it has been reported that such aid by a helper serves to improve the reproductive success of the pair[1,2,3]. However, such a trait, whereby an individual does not breed but helps another individual in its breeding and derives no benefit from this activity seemed evolutionarily unsustainable. For long, this trait was a source of torment to Darwinian evolutionists. The resolution of this apparent evolutionary paradox is one of the prime reasons why scientists have focused considerable attention on the study of cooperative breeding. The review by S. T. Emlen[3] provides details regarding the adaptive significance of cooperative breeding as described hereinafter.

The theory of kin selection by W. D. Hamilton[4] offered a solution to this challenging question. Hamilton's theory explains that in addition to helpers, kin-related individuals could also exhibit traits of helping behavior; further, it suggests that if a kin individual's reproductive success increases through this helping behavior, the trait will be passed on from generation to generation. In this manner, the population will eventually be comprised of individuals that have the same traits. Furthermore, instead of enhancing their own reproductive success (i.e., the total number of offspring produced during an individual's lifetime; also termed "fitness"), if individuals enhance the reproductive success of other kin individuals, they can thereby enhance the ratio of individuals in the population with similar genes as themselves (referred to as "heightening genetic contribution"). The evolutionary benefit gained in this manner is termed the "indirect benefit (of helping behavior)."

This interpretation of the adaptive significance of cooperative breeding through the theory of kin selection stimulated the interest of scientists. Applying this theory, it could be predicted that the indirect benefit derived by the kin-related helper is larger than the benefit that it would gain had it reproduced by itself. It is necessary to determine those benefits in order to verify the hypothesis. Therefore, the indirect benefits of helping behavior were determined in the case of several species. These were then compared with the benefits that helpers would derive if they reproduced themselves. However, it was found that in most species, an individual achieved a greater genetic contribution to descendants (greater fitness) when it reproduced by itself than it would if it had helped others. Expressed solely in terms of indirect benefits, the theory was untenable.

Consequently, scientists directed greater attention toward identifying ecological factors rather than genetic factors. The possible direct ecological benefits to the helper are as follows:

(1) Group size increases when helpers enhance the reproductive success of breeding individuals. This facilitates the maintenance and expansion of territory. Through this practice, the helper itself can secure resources such as food and shelter within the territory and enhance its survival rate. Additionally, there is an increased possibility that a part of the expanded territory may be portioned to it or that it will inherit the currently existing territory. This case is based on the premise that territory is a scarce and limited resource and that if the individual were to disperse from the natal territory, it would find survival difficult and would be unable to breed.

(2) Through its helping behavior, an individual gains some experience and knowledge in nestling care, which it can use to enhance its own reproductive success in future.

(3) It can nurture a helper that will assist its own reproduction in future.

(4) When the breeding female is not a kin-related individual, the helper could be accepted as a breeding mate.

Confirmation of the existence of helpers

Within the family Vangidae, individuals other than the pair assisting in brood care have been observed among several species, the Rufous Vanga[5], the Chabert's Vanga[5], the White-headed Vanga[6], and the Sickle-billed Vanga[7]; however, no detailed records existed. As described in Chapter 1, in 1991 at the

Ampijoroa research forest in western Madagascar, Dr Satoshi Yamagishi and his colleague observed that pairs of Rufous Vangas were accompanied by adult males (black throat) and yearling males (spotted throat); further, those auxiliary males assisted in provisioning the chicks[8]. However, the origin of the auxiliary helpers, the effectiveness of the assistance, etc., remained unknown. Since it is a known fact that in several cooperatively breeding species, sons stay at home and assist the parents, we could predict the origin of the yearling male to a certain extent. However, questions persist with regard to the adult helper male. Is it a grown-up son? Or, as seen in the case of the Dunnock, do two unrelated males form a trio along with a female[9]? This was the starting point for our project, "Social Evolution of the Vangidae," which was launched in 1994.

Daughters leave, sons remain

In most species of birds and mammals, once the young have reached a certain level of maturity, they disperse from home. This is termed natal dispersal. There is a sexual difference in the timing and distance of dispersal. Among birds, females disperse further, whereas males usually establish a territory in a location adjacent to their natal territory. Among cooperatively breeding bird species, it has been observed in several cases that females leave home at the age of one and acquire a breeding mate, whereas males stay at home and later form a new territory near the natal territory. The Rufous Vanga is a typical example of such a case.

Table 6-1 presents the situations of one-year-old individuals during the breeding season; survival of these individuals was confirmed up to one year after fledging[10]. By the beginning of the breeding season, several females had dispersed beyond the study area. Only 10 remained, seven of which acquired a breeding mate after dispersing from the natal territory. During the six-year study period, only three females stayed at home; however, they seldom associated with other members. In contrast, most of the males stayed at home and no individuals bred at the age of one. Exceptional cases were observed: four males dispersed from their natal territories, three of which became helpers after joining different territory groups, and the fourth became a floater.

Why do yearling males not disperse? One reason is that they are sexually immature and sterile. In the course of our 1998 research, we extracted the testes of three captured individuals-one dominant (breeding) male over two years old and two yearling helpers. These individuals belonged to the same group and were building a nest when captured. We captured them during that particular period because we believed that they were all physiologically capable of

Table 6-1. Status of surviving yearling Rufous Vangas (Eguchi et al. 2002)[10)]

	Philopatric			Dispersed	
	Helper	Non-helper	Unknown	Bred	Non-breeder
Male	13	13	14	0	4[1]
Female	0	3	0	7	18[2]

[1] Three moved to non-parental territories and stayed there; one became a floater.
[2] All emigrated from the study area by the beginning of the breeding season.

Table 6-2. Testes size of the Rufous Vanga (Yamagishi et al. 2002[11)])

Individuals	Age	Body mass (g)	Tarsus (mm)	Combined testes mass (mg)	Right testicular volume (mm^3)
Breeder	Over two-year-old	37.6	24.3	171.5	75.4
Helper	One-year-old	39.4	24.4	60.8	38.5
Helper	One-year-old	39.8	23.8	14.5	13.1

breeding and that their testes were sufficiently mature. However, despite the large body size of the yearling helpers, their testes were much smaller than those of the over two-year-old male (Table 6-2)[11]. A microscopic examination of the cross section of the testes revealed that those of the breeding (dominant) male contained adequate quantities of sperm. However, in the testes of the two yearling helpers, sperm was scarce, and it was evident that these helpers were immature (Plate II-6). Although the number of examined individuals is limited, it does appear that yearling males of the Rufous Vanga are sexually immature. In contrast, although their testes have not been examined, helpers over two years old were sometimes observed attempting to copulate with pair females when the breeding (dominant) male was absent[11]. Therefore, it appears that helpers over the age of two are sexually mature. If yearling males are sexually immature, maintaining territories, if acquired, would not benefit them in any way. It is more advantageous for helpers to receive benefits (security and food resources) by staying at home until they reach maturity and to collaboratively defend the territories of their parent(s) than to address several challenges and interactions independently while maintaining their own territory.

What practice do yearling males follow? As shown in Table 6-1, out of 44 yearling males, 29 remained in the study area the following year. Nine of these acquired independence, territories, and breeding mates; however, 20 individuals continued as helpers. Of these 20 male individuals, 16 were in their natal territories and four became helpers in different territories (Fig 3-9). Clearly, it is not easy to acquire territories even after growing fertile.

The strong natal philopatry of males produces a situation wherein most of the

Table 6-3. Pattern of independence (Eguchi et al. 2002)[10)]

	Two years old	Three years old or older
Total	12	9
Replaced in the same territory	0	1
Adjacent territory		
Newly established	8	6
Takeover	1	0
Spaced one territory distance		
Newly established	2	0
Takeover	0	1
Floater or not stable	1	1

helpers accompanying a pair are their sons. Although the proportion of pairs that are accompanied by helpers fluctuates annually, it accounts for approximately one third (24~43%) the entire breeding population. Among the helpers, 33~62% are males over two years of age (adult males). Two-thirds of the groups with helpers include at least one helper over two years of age (Plate II-7, Table 3-3). Why do these individuals remain helpers even after reaching sexual maturity?

As observed in the case of the family of the "pink-stripe banded individual" in Chapter 3, males disperse solitarily and acquire independence, and most of them establish new territories adjacent to their natal territory where they spent time as helpers (Table 6-3)[10]. In other words, they establish territories at a site that is partitioned off from the natal territory or in the space between neighboring territories. We observed only one case wherein a subordinate helper acquired the female mate and the dominant status after the breeding (dominant) male disappeared. There were only two cases in which an invader male usurped a territory. Even in the case of usurping, the usurped territories were limited to those close to their natal territories (Table 6-3).

The reason why males stay at home can be explained in terms of the following facts. Firstly, floater individuals are rare. This indicates that if an individual male does not belong to any group, it will most likely experience difficulties in its daily quest for survival. As described in Chapter 5, food resources do not seem scarce. The study area consists of a series of forests (Plate I-4), suitable habitat is not distributed like islands surrounded by undesirable habitat, and preferred habitats do not appear to be scarce. While it is possible that the site of this forest is saturated with territories, the territories do not seem too densely packed. If floater individuals move around, they inevitably intrude into a territory and are attacked by its occupants. Therefore, floaters may find it difficult to engage in feeding without encountering harassment or interruptions. It is considered challenging for a lone individual to combat harassment by the more numerous territory owners. If a male formed a pair with another individual, then together they might be able to resist the attacks of territory owners. However, an immature male is not even capable of acquiring a female. Therefore, as mentioned above, staying within the natal territory and defending the territory collaboratively is the optimal tactic. Moreover, as is evident in their pattern following independence, males can enhance their opportunity of acquiring a territory by staying at home. It is unlikely that the territories are completely saturated. During the six years since 1994, four territories disappeared and eight new territories were formed in a specific quadrat of the study area. Eventually, four territories were added. Thus, it appears that the territories were not fully saturated. If these immature males persist, space to establish a new territory is available between the borders of adjacent territories. Yet, no solitary yearling individuals are found. Therefore, we believe that this situation is not one wherein dispersal is impossible due to territory saturation and lack of space in which to found a new one; rather, yearling males delay their dispersal to derive the benefit that can be accrued by staying at the natal territory (the "Benefits-of-Philopatry hypothesis"[12]).

The possibility of producing offspring through copulation with the stepmother is heightened if the yearling male stays at home. Breeding females transfer between territories more frequently than do males (Figs. 3-7 and 3-8). In eight territories that were located in the midsection quadrat of the study area, the average period for which breeding individuals inhabited a single territory was 5.1 years among males and 2.4 years among females. When emigration or immigration occurs, there is an increased probability that the new female and helper(s) are unrelated; therefore, helpers are able to produce offspring by copulation with the stepmother. In fact, in 1997, a subordinate helper in Group-TX was observed copulating with the pair female; the result of DNA parentage determination also implied the probability of fertilization by the male helper[11]. During the same year in Group-L as well, a helper copulated with a stepmother that had newly immigrated that year. This individual was still a helper in Group-L during the breeding season of 1999. Having hatched in 1994, it held the status of a helper for the longest duration (for four years since it grew into a black-throat male), in a manner similar to that of another helper individual in Group-H. There is no indication of these helpers becoming independent, I wonder if they chose to continue a triangular relationship involving their father? In this manner, the helper retains its chances of reproducing and can also remain at home.

Helpers enhance their opportunities of reproduction by staying at home. Whatever the manner, once they acquire the status of a breeding male, their position is relatively stable. Of the 13 individuals leg banded in 1994, eight were still defending their breeding male position six years later. One other individual had defended its position for five years. Generally, tropical birds have a long

郵便はがき

料金受取人払

左京局
承認
1134

差出有効期限
平成18年
12月31日まで

606-8790

（受取人）
京都市左京区吉田河原町15-9　京大会館内

京都大学学術出版会
読者カード係　行

■ご購読ありがとうございます。このカードは図書目録・新刊ご案内のほか、編集上の資料とさせていただきます。お手数ですが裏面にご記入の上、切手を貼らずにご投函ください。

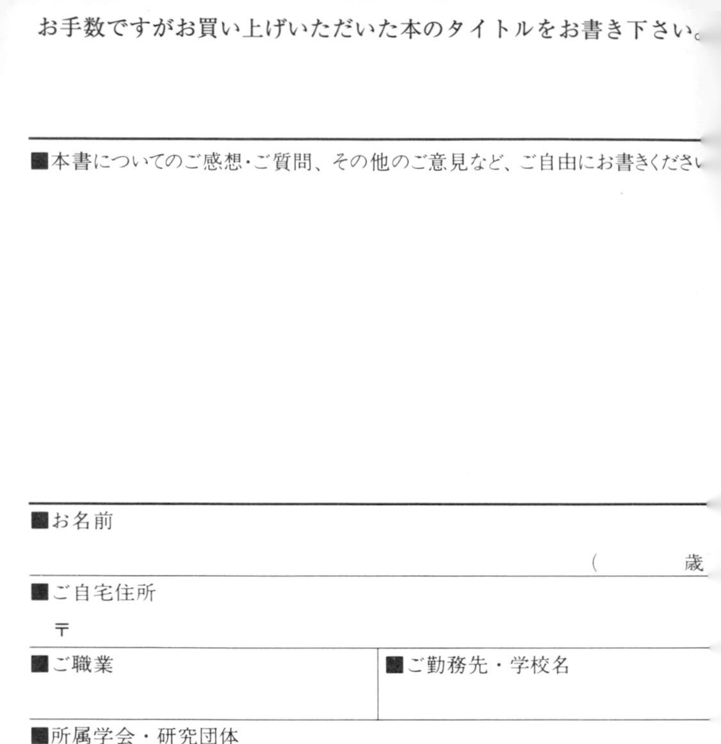

お手数ですがお買い上げいただいた本のタイトルをお書き下さい。

■本書についてのご感想・ご質問、その他のご意見など、ご自由にお書きください

■お名前

（　　　歳

■ご自宅住所

〒

■ご職業

■ご勤務先・学校名

■所属学会・研究団体

●ご購入の動機

A．店頭で現物をみて　B．新聞広告（紙名　　）
C．雑誌広告（誌名　　）　D．小会図書目録
E．小会からの新刊案内（DM）　F．書評（　　）
G．人にすすめられた　H．テキスト　I．その他

●ご購入書店名

都道府県　市区町　書店

京都大学学術出版会 TEL（075）761-618
FAX（075）761-619

lifespan. Is this a kind of lesson for humans, suggesting that one should perform ascetic training while still young like the helpers? Well, there are also some helper individuals that remain at home and can acquire territories without undergoing any training (assisting).

Is helping essential?

Helpers assist breeding pairs, and the assistance they provide ranges from territory defense and anti-predator mobbing behavior to chick provisioning. Figure 6-1 shows the degree of helper participation in these activities[10].

Helpers participated in approximately 60% of the cases where mobbing behavior was observed and in approximately 80% of the instances of territorial defense. Whenever predators appeared or territorial intrusion occurred, almost all individuals in the vicinity seemed to participate, and the tendency for a particular individual within the group to lead the defense behavior was not observed. This implies that every individual within the group shoulders an equal amount of risk.

During chick provisioning, helpers contributed approximately 25% of the total feeding frequency, and this degree of contribution is almost equal to that of the breeding female (Fig 6-1). However, not all helpers assist in provisioning. Among the 26 yearling males examined, 13 individuals participated in provisioning, whereas the remaining 13 never visited the nest (Table 6-1). An examination of every group with helpers revealed that the helper never participated in provisioning the chicks in approximately 30% of these groups. To date, studies of cooperative breeding have concluded that every helper contributes almost equally[12]. However, among the Rufous Vangas, there exists a considerable difference in the degree of contribution between individuals that assist and those that do not. For example, even among helpers within the same group, an individual that frequently provisioned the chicks participated in over 40% of all feedings, whereas the contribution of the other helper was nil. Ultimately, this implies that more than 30% of helpers are permitted to remain within the parent's territory despite the fact that they do not contribute any assistance in provisioning the chicks.

When we consider cooperative breeding, we tend to regard it as a matter of course that every helper assists. Usually, we accept the simple formula that a helper is directly equivalent to an assistant. However, 30% of helpers, or sometimes almost 50% (depending on the year), do not contribute at all, at least with respect to provisioning the chicks. Despite this fact, they are allowed to stay within their natal territory, and in this respect, the Rufous Vanga is perhaps

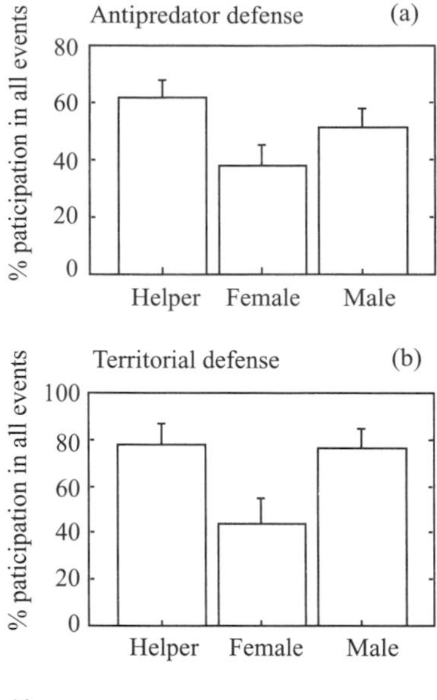

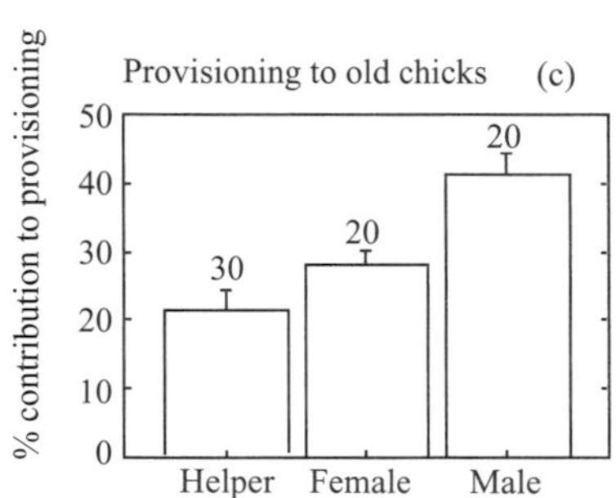

Fig. 6-1 Contribution of group members to: (a) anti-predator defense (27 groups), (b) territorial defense (20 groups), and (c) provisioning of chicks after Day 8. Data for helpers are combined for all helpers in the same group. Vertical bars indicate SE. Values above error bars are numbers of individuals observed, (modified from Eguchi et al. (2002)[10])

unique among cooperatively breeding birds. The degree of contribution of some individuals is extremely high; therefore, when we consider the cooperative breeding of the Rufous Vanga, it is important to take into account the existence of "non-assisting individuals." When such complicated social situations arise, it is necessary to clearly distinguish between "non-assisting helpers" and "assisting helpers."

As regards the feeding contribution of helpers, a marginal difference exists between yearling males and adult individuals. Adult helpers feed the chicks when they are still young, whereas the degree of contribution by yearling helpers at this stage is low (Fig 6-2). However, when the chicks grow up, the degree of contribution of yearling helpers is almost equal to that of adult helpers.

The frequency of provisioning activity differs greatly among helper individuals. It has occasionally been suggested that the presence or absence of

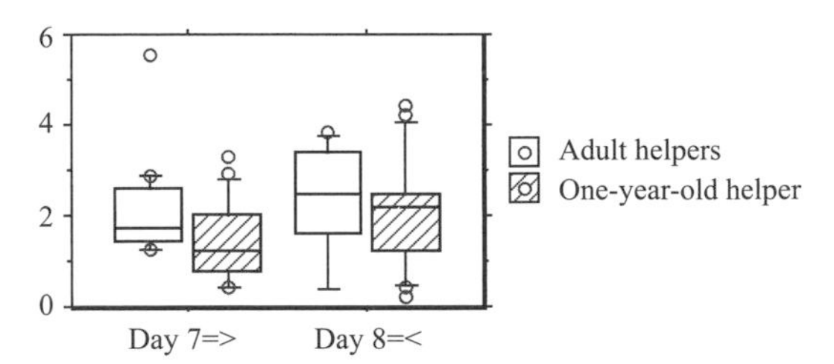

Fig. 6-2 A comparison of provisioning frequency between adult helpers and one-year-old helpers. A difference is significant for the provisioning to young chicks.

paternity is one reason for the existence of such differences in provisioning frequency. For example, in one instance, two fertile males formed a trio with one female. A dominance hierarchy was formed between the two males, and the subordinate male was deprived of its mating opportunity by the dominant individual. Among the Dunnock, subordinate males seek the opportunity to mate with the female. If copulation succeeds, then the degree of nestling care by the subordinate male increases[9]. Among the White-browed Scrub Wren *Sericornis f. frontalis*[13] too, subordinate helper males that copulate with an unrelated female increase their frequency of provisioning chicks. In the case of the Dunnock, in instances where the opportunity for copulation is low, the subordinate male seldom feeds the newly hatched chicks. Considered together, these observations indicate the following. If the subordinate male is unrelated to the female, copulation can be undertaken without any concern regarding an incestuous relationship. Thus, if the possibility to produce offspring is increased, the subordinate male feeds the young enthusiastically in order to enhance the survival rate of its own progeny. Conversely, if the female is the helper's mother, copulation does not occur. Further, even if unrelated, the opportunity to copulate does not arise; hence, the chicks that hatch are not those of the helper, and accordingly, its feeding assistance is not so enthusiastic.

The same trend is expected in sterile yearling males. By demonstrating their provisioning ability to the female, yearling males may seek to maximize the possibility that they will be accepted (as a provisioning breeding male) when they become mature the following year. In such a case, wherein it is not related to the female, it will actively provision the young although it is only a yearling male. However, contrary to predictions, no increase in provisioning frequency, which would have taken place had the breeding female been an unrelated individual, was observed. For example, a comparison of the provisioning frequency of groups that included yearling males and adult males revealed that

Table 6-4. The change in frequency of attacks by breeding males against helper males in relation to the stage of the breeding season (data from 1995 to 1997). Figures indicate the number of attacks/observation time (min.). (Eguchi et al. 2002)[10)]

Breeding stage	One-year-old helpers	Adult helpers
Nest building	4/941	108/6769
Egg-laying	0/1196	51/5749
Incubation	0/4385	7/8018
Nestling	0/16514	1/8595

the provisioning frequency toward unrelated females was 1.39 ± 1.67 times (n=9), whereas in cases wherein helpers were related to the female, the provisioning frequency was 1.88 ± 1.50 times (n=17); no significant difference existed between them. Further, a comparison of groups that included only adult males revealed that the provisioning frequency toward unrelated females and related females was 2.48 ± 1.66 times (n=4) and 2.25 ± 1.59 times (n=5), respectively. Again, no significant difference exists between them. Moreover, in 1997, the helper of Group-TX, which was judged to have fathered some of the young in that nest on the basis of parent-offspring relatedness, did not increase its provisioning rate. However, quantitative data is scarce, and it is necessary to confirm paternity in several individuals and to compare the provisioning frequencies before arriving at a definite conclusion. Thus, a clear explanation of the existence of "assisting helpers" and "non-assisting helpers" remains elusive.

Helpers also attempt to assist by carrying nest materials and by participating in incubation. However, their contribution rate in these activities is low, e.g., carrying nest materials: 4% in yearling males and 5% in adult helper males; incubation behavior: 1% in yearling males and 10% in adult helper males. The degree of contribution is low because the breeding male expels helpers that approach the nest, irrespective of whether they are carrying nest materials or intending to assist in incubation. At the onset of the breeding period, the breeding male often attacks helpers (particularly adult helpers) (Table 6-4), and it would appear that the breeding male is rejecting the helper's assistance. However, as the breeding stage advances and enters the nestling-care period, the breeding male seldom attacks the helpers. Attacks directed toward yearling males seldom occur; instead, attacks are directed toward adult male helpers. The frequency of such attacks remains high until the egg-laying period, which suggests that these attacks are motivated by mate-guarding instincts. Yet, as noted in Chapter 5, pair males do not appear to be particularly vigilant regarding mate guarding.

We occasionally noted certain behaviors that led us to question whether the nest-material-carrying behavior of helpers was truly a form of assistance. We occasionally observed helpers carrying nest materials to a location other than the nest site. Such behavior might indicate the independence of the carrier. However, this behavior did not last for long; the individual stopped this behavior as aberrantly as it had begun and resumed assisting the breeding pair. In several instances, nest materials were carried in small amounts, and the carrying of materials to a different place persisted for only about one day.

Will filial piety reap a reward?

It is evident that assisting helper individuals work hard. However, does this assistance ultimately enhance reproductive success? In several species, it has been reported that helping behavior improves reproductive success. However, this apparently satisfying conclusion was derived from a simple correlation between the number of helpers and the reproductive success of the pair; the influence of other factors was not controlled for. Consequently, this conclusion has been criticized as confusing cause and effect, and it has been stated that the results have been interpreted incorrectly[3]. For example, even if helpers do not assist, a simple effect of group size can enhance the efficiency of territorial defense, thereby increasing the survival rate of the group members and chicks. Moreover, in a high-quality territory, chick survival is high, several chicks remain within the territory, and accordingly, the number of helpers increases. This implies that the abundance of helpers is the result and not the cause of a high reproductive capacity.

In the case of the Rufous Vanga, pairs accompanied by helpers were more successful in breeding; 45% of such pairs (31/69) were successful, as compared with a 30% (35/118) success rate for pairs without a helper. Pairs with a helper also fledged more chicks (1.19 $\pm$ 1.53 chicks per nest, n=60) than did pairs without a helper (0.62 $\pm$ 1.11 chicks per nest, n=117). Moreover, a comparison between pairs that received assistance in provisioning by helpers and those that did not revealed that the former enjoyed a higher reproductive success (the success rates were 84%, i.e., 21/25 pairs and 55%, i.e., 40/73 pairs, respectively) and successfully fledged a greater number of chicks (2.08 $\pm$ 1.41 chicks per nest, n=25 vs. 1.19 $\pm$ 1.33 chicks per nest, n=72). Thus, in both examples, we observe that pairs accompanied by helpers experience greater reproductive success. Such findings are the usual outcome of conventional studies, and they form the basis of the conclusion that helping behavior or the presence of helpers enhances reproductive success. However, it is not possible to deduce causal relationships on the basis of such comparisons of mean values and correlations. Therefore, we performed a further analysis, considering only pairs that were accompanied by helpers and differentiating between the above-mentioned categories of "assisting helper" and "non-assisting helper." This analysis revealed no significant difference between the reproductive success of pairs with assisting helpers and those with non-assisting helpers: the proportion of successful pairs was 84% (21/24 pairs) and 75% (6/8 pairs), respectively; the number of fledged chicks was 2.08 ± 1.41 chicks per nest (n=25) and 1.88 ± 1.46 chicks per nest (n=8), respectively. This finding strongly suggests that there exists no difference in reproductive success between pairs with assisting helpers and those with non-assisting helpers and that helping behavior, thus, does not actually enhance the reproductive success of pairs[10].

It is possible that the higher reproductive success observed in pairs with a helper might be due to factors other than helping behavior. During the six years of this study, we found that the number of helpers attending each pair changes yearly and that a helper may be present or absent. A comparison of those cases wherein the combination of pair members remained unchanged and only the presence or absence of helpers differed revealed that the presence of helpers did not affect a pair's reproductive success. In other words, in the 18 cases examined, the number of chicks fledged in nests during the years when a helper was present (1.55 ± 1.15 chicks per nest) was not significantly different from the number of chicks fledged during years when no helper was present (1.39 ± 1.72 chicks per nest).

Secondly, the apparent correlation between the presence of a helper and reproductive success could be the outcome of high reproductive productivity within a specific territory. We therefore investigated this possibility as well. It is difficult to directly evaluate the quality of a territory; hence, we employ another indicator here. During the research period, there were some pairs that did not have a helper and others that were accompanied by a helper every year. We selected data on pairs for which breeding records were taken for more than two years. We then divided these pairs into those that were never accompanied by a helper and those that were accompanied by a helper for at least two years. Finally, we compared the number of chicks that had fledged during the years when both these groups did not have a helper. Results indicated that the pair accompanied by the helper for at least two years fledged more chicks (1.47 ± 1.15 chicks, n=18) than the pairs that were never accompanied by a helper (0.52 ± 1.15 chicks, n=12). This implies that the reproductive productivity of a given pair is equally successful, irrespective of whether or not they are attended by a helper[10].

Such results on the Rufous Vanga contradict the prevailing conventional viewpoint that helping behavior invariably serves to enhance the reproductive success of breeding pairs[1,2,3]. An almost identical pattern has reportedly been observed among the White-browed Scrub Wren[14] and the Laughing Kookaburra *Dacelo novaeguineae*[15]. The Rufous Vanga is the third bird species in which it has been demonstrated that helping behavior has no positive effect. Why is it that helping behavior (in this case, chick provisioning behavior) does not lead to increased reproductive success? As indicated in Chapter 5, the reproductive success of the Rufous Vanga is basically low, and the predation of eggs and chicks is considered to be a major causal factor. In fact, apart from predation-related losses, chick mortality during the nestling period is extremely low. This implies that chicks seldom die due to starvation; therefore, even if the feeding frequency increases, it does not directly increase the survival rate of the chicks. However, if the predation rate is high, the survival rate of chicks could be increased by ensuring that chicks fledge as soon as possible, and thereby, they spend a minimal amount of time in the nest. If the increase in the quantity of food through additional feedings serves to accelerate chick development, then reproductive success will improve. However, a comparison of the weight of 10-16-day-old chicks of pairs with a helper and those without indicated that there was no significant difference (chick body weight 30.2 ± 1.7 g, n=17 and 30.8 ± 2.4 g, n=21, respectively); further, there was no difference in the nestling-care period (15.7 ± 1.18 days, n = 18 and 15. 6 ± 1. 26 days, n=28, respectively).

What then are the causal factors of the difference in reproductive success? Due to the limited amount of data available, we cannot confirm whether parental quality affects reproductive success. The quality of the territory may also play an important role. However, food shortage is not a significant factor in reproductive failure; therefore, in this case, food resources are not regarded to be a limiting factor. Predation is considered to be the main cause of reproductive failure; accordingly, reproductive success may be higher in a safe territory. Unfortunately, we did not record data on factors such as predation pressure, the frequency of appearance of predators, etc., which would facilitate a comparison of predation pressure between territories. Therefore, we are presently unable to confirm or refute this hypothesis.

Is it true that even though some helpers provide high levels of assistance, they do not benefit the breeding pair in any way? With regard to reproductive success at least, there is no improvement. There are some reports that helpers reduce the quantity of labor expended by pair members by shouldering a part of the provisioning activities, thereby enhancing the survival of pair members. The Rufous Vanga has a long life span, and data recorded during the course of our six-year study period are insufficient to facilitate a comparison of the survival rates among breeding individuals. However, among pairs accompanied by a helper, a decrease in the feeding frequency by the pair female was evident during the latter part of the nestling period when the female was required to

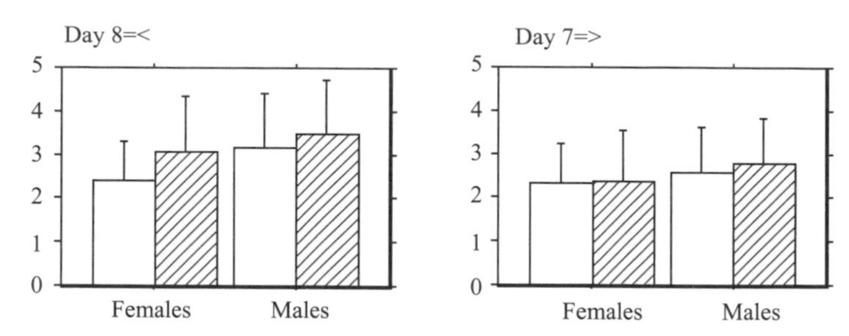

Fig. 6-3 Effect of helping on reciepients' provisioning rate. The provisioning frequency of females to old chicks is significantly lower in pairs with helpers (open columns) than in unaided pairs (shaded columns).

work the hardest (Fig 6-3). This reduction in the amount of labor might serve to increase the female's survival rate. However, females often change territories; therefore, instances wherein the female is not related to the helper also occur. In such a situation, even if the helper increased the female's survival rate, the helping behavior does not necessarily heighten the genetic contribution of the helper to the next generation.

One good turn deserves another

The helper might derive direct benefits from its provisioning behavior. First, providing food to the chicks might serve as the "rent" that permits them to remain in the pair's territory. In such a situation, we might expect lazy helper individuals to be expelled by the breeding pair. To evaluate this possibility, we first excluded individuals that became independent. We then compared the number of yearling individuals that migrated from the natal territory to another territory or disappeared in the following year. Out of 11 yearling males that provisioned chicks, seven remained in the same territory the following year and four dispersed from the territory. In comparison, out of 10 yearling males that did not exhibit provisioning behavior, eight remained in the territory and two dispersed. There is no significant difference between these two cases, and it appears that laziness does not result in expulsion. Attacks on yearling males were rare. However, at the beginning of the breeding period, attacks directed toward adult male helpers were frequent. Such attacks were directed toward individuals that visited the nest to participate in nest building and incubation. The breeding (dominant) male appears to reject the assistance proffered by these helpers. In brief, the helpers are not attacked even if they do not assist.

It is possible that yearling males help not because they want to stay in the territory but because they can improve their nestling-care techniques through the practice gained by provisioning the young, and consequently be able to heighten their reproductive opportunities after becoming independent. Thus, as opposed to the idea of "rent," we can infer that the more enthusiastic the helper, the higher the possibility of its leaving the territory. However, our data does not support this prediction either. Excluding those individuals that disappeared, we compared the yearling males that stayed in the territory the following year and those that established a new territory within the study area. One of the eight males that assisted and four of the 12 males that did not assist became independent. Although it would appear that relatively more non-assisting individuals than assistance-providing helpers became independent, this difference is actually non-significant. It appears that helping does not enhance a helper's opportunity of becoming independent. Unfortunately, the data we collected during our six-year research period are insufficient to clarify whether a helper's experience of providing care to nestlings subsequently enhances its own reproductive success.

When a helper becomes independent, it leaves the territory on its own. It is neither accompanied by a future mate nor by its own helper. When feeding duties are alternated at the nest, the individual that is already at the nest leaves first, and then the food bringer enters the nest. In the case of the White winged Chough *Corcorax melanorhamphos*[16] and the Arabian Babbler *Turdoides squamiceps*[17], the provisioning individual notify the other members about the feeding behavior.

However, no such a behavior was seen among the Rufous Vanga.

Evidence suggests that the breeding pair does not expel helpers, irrespective of whether or not they actively participate in provisioning the chicks. Helpers also participate in territory defense and anti-predator mobbing. The extent to which each individual contributes toward these activities has not been examined; however, on the basis of our observations, all individuals in the vicinity respond to any incident that occurs. The helper also derives direct benefits from territorial defense and mobbing. Perhaps that is why the helper participates in these activities whenever the opportunity arises. It is advantageous when a large number of individuals involve themselves in territorial defense and mobbing behavior; therefore, it is considered that the larger the number of individuals protecting the territory, the greater is the advantage derived by the occupants. In such a situation, the breeding pair would not expel the helper from the territory even if that helper does not actively provision the chicks.

Is this situation maladaptive?

In 1987, I. G. Jamieson and J. L. Craig proposed a theory stating that the provisioning behavior of a helper toward chicks of other birds is in fact a maladaptive trait induced as a conditioned response to the begging behavior of chicks[18]. This was swiftly rebutted by D. J. Ligon and P. B. Stacey. In 1989, Ligon and Stacey proposed that while provisioning by non-parental individuals might have originally been induced by the begging behavior of chicks, as asserted by Jamieson and Craig, at present, the behavior of helper individuals in a cooperatively breeding species has a different adaptive significance from its origin. Ligon and Stacey concluded that it is not appropriate to surmise that all provisioning behaviors are maladaptive and are a direct response to the begging behavior of chicks[19]. They also noted that each species experiences different selection pressures on the provisioning behavior of helpers and that instances of behaviors that do not have an adaptive function will be present in some species. Accordingly, they regarded cases wherein the provisioning behavior of the helper is useful to neither the breeding pair nor the helper itself as unremarkable.

In the case of the Rufous Vanga, although marked differences exist in the degree of provisioning undertaken by individual helpers, this difference does not appear to affect either the reproductive success and survival rate of the pair or the benefits accrued by helpers. Furthermore, the breeding pair does not expel the helper on the basis of the degree of assistance that it provides. Moreover, depending on the breeding stage, they sometimes do not accept the helper's assistance. These observations imply that helping behavior among the Rufous Vangas has not evolved to a stage wherein it has an adaptive significance. I am presently unable to accept the hypothesis proposed by Ligon and Stacey; however, from the perspective that several individual helpers remain at home without helping, their hypothesis might yet be proven valid.

Among certain cooperatively breeding species, high costs exacted due to helping behavior have been documented[20,21]. However, in the case of the Rufous Vanga, there is no difference in the proportion of yearling males surviving to the following years between individuals that assist (78.9%, n=19) and those that did not assist (71.4%, n=14). If the cost is actually low, helping behavior might become established and spread through the population even if the benefit is minimal. If the behavior itself does not involve a large cost, it will be maintained in the population even if it holds no adaptive significance. Among the Rufous Vangas, no large costs related to the survival of provisioning behavior are evident.

It is clear that provisioning by helpers is not beneficial in terms of enhancing reproductive success. However, in order to conclude that it is non-adaptive, it is first necessary to obtain data on the direct benefits that the helper itself might derive. However, such information is scarce. Furthermore, data on each individual's contribution toward territory defense and mobbing behavior is also necessary in order to resolve this question.

Conclusion

Approximately 30% of the breeding pairs are accompanied by a yearling male helper or a male helper that is over two years old. A majority of these are sons of the breeding pair that have remained within the natal territory. These helpers participate in activities like anti-predator mobbing, territory defense, provisioning chicks, etc. However, the degree of contribution toward provisioning the chicks differs greatly between individuals, and approximately half of these individuals play no part whatsoever in chick provisioning. It is believed that helpers remain in the natal territory in order to enhance their possibility of acquiring that territory and of copulating with the stepmother. In pairs accompanied by a helper, reproductive success is high; yet, a pair's reproductive success is not affected by whether or not the helper assists in provisioning the chicks. Likewise, provisioning by the helper neither accelerates the growth rate of the chicks nor shortens the nestling period. Thus, it does not lower the risk of predation. Some pairs often breed with the assistance of a helper, and even during those years when the helper is absent, the reproductive success of such pairs is higher than that of pairs that are seldom accompanied by a helper. These facts collectively indicate that the presence or absence of a helper, and whether the helper actually assists or not, does not affect the reproductive success of the pair. Such characteristics imply that the Rufous Vanga is unique among the species of cooperatively breeding birds.

References

1 Brown, J. L. (1987) *Helping and Communal Breeding in Birds*. Princeton University Press, Princeton.
2 Stacey, P. B. & Koenig, W. D. (1990) *Cooperative Breeding in Birds: Long-term Studies of Ecology and Behaviour*. Cambridge University Press, Cambridge.
3 Emlen, S. T. (1991) Evolution of cooperative breeding in birds and mammals. *Behavioural Ecology: An Evolutionary Approach*, 3rd ed, (eds. J. R. Krebs & N. B. Davies), pp. 301-337, Blackwell, Oxford.
4 Hamilton, W. D. (1964) The genetical evolution of social behaviour. I. *Journal of Theoretical Biology*, 7: 1-16.

5 Appert, O. (1970) Zur Biologie der Vangawurger (Vangidae) sudwest Madagaskars. *Ornithologische Beobachter*, 67: 101-133.
6 Nakamura, M., Yamagishi, S., & Nishiumi, I. (2001) Cooperative breeding of the White-headed Vanga *Leptopterus viridis*, an endemic species in Madagascar. *Journal of Yamashina Institute for Ornithology*, 33: 1-14.
7 Nakamura, M., Yamagishi, S., & Okamiya, T. (2001) Breeding ecology of the Sickle-billed Vanga *Falculea palliata*, which is endemic to Madagascar. *Ecological Radiation of Madagascan Endemic Vertebrates* (eds. S. Yamagishi & A. Mori), pp. 48-52, Kyoto University, Kyoto.
8 Yamagishi, S., Urano, E., & Eguchi, K. (1995) Group composition and contributions to breeding by Rufous Vangas *Schetba rufa. Ibis*, 137: 157-161.
9 Davies, N. B. (1992) *Dunnock Behaviour and Social Evolution*. Oxford University Press, Oxford.
10 Eguchi, K., Yamagishi, S., Asai, S., Nagata, H., & Hino, T. (2002) Helping does not enhance reproductive success of cooperatively breeding Rufous Vanga in Madagascar. *Journal of Animal Ecology*, 71: 123-130.
11 Yamagishi, S., Asai, S., Eguchi, K., & Wada, M. (2002) Spotted-throat individuals of Rufous Vanga *Schetba rufa* are yearling males and presumably sterile. *Ornithological Science*, 1: 95-100.
12 Stacey, P. B. & Ligon, J. D. (1991) The benefits-of-philopatry hypothesis for the evolution of cooperative breeding: variation in territory quality and group size effects. *American Naturalist*, 137: 831-846.
13 Magrath, R. D. & Whittingham, L. A. (1997) Subordinate males are more likely to help if unrelated to the breeding female in cooperatively breeding white-browed scrubwrens. *Behavioral Ecology and Sociobiology*, 41: 185-192.
14 Magrath, R. D. & Yezerinac, S. M. (1997) Facultative helping does not influence reproductive success or survival in cooperatively breeding white-browed scrubwrens. *Journal of Animal Ecology*, 66: 658-670.
15 Legge, S. (2000) The effect of helpers on reproductive success in the laughing kookaburra. *Journal of Animal Ecology*, 69: 714-724.
16 Boland, C. R. J., Heinsohn, R., & Cockburn, A. (1997) Deception by helpers in cooperatively breeding white-winged choughs and its experimental manipulation. *Behavioral Ecology and Sociobiology*, 41: 251-256.
17 Zahavi, A. (1990) Arabian babblers: the quest for social status in a cooperative breeder. *Cooperative Breeding in Birds: Long-term Studies of Ecology and Behaviour* (eds. P. B. Stacey & W. D. Koenig), pp. 105-130, Cambridge University Press, Cambridge.
18 Jamieson, I. G. & Craig, J. L. (1987) Critique of helping behaviour in birds: a departure from functional explanations. *Perspectives in Ethology*, vol. 7, (eds. P. Bateson & P. Klopfer), 79-98, Plenum Press, New York.
19 Ligon, J. D. & Stacey, P. B. (1989) On the significance of helping behaviour in birds. *Auk*, 106: 700-705.
20 Heinsohn, R. & Cockburn, A. (1994) Helping is costly to young birds in cooperatively breeding white-winged choughs. *Proceedings of the Royal Society of London*, B256: 293-298.
21 Heinsohn, R. & Legge, S. (1999) The cost of helping. *Trends in Ecology and Evolution*, 14: 53-57.

Sons or daughters?

Shigeki Asai

Why is the number of males and females almost equal among several animals?

This section explores the factors underpinning the sex ratio (the number of males:the number of females) observed among the Rufous Vangas. As described later, the sex ratio of the Rufous Vanga is strongly male-biased. In numerous animals, the sex ratio is almost 1:1; therefore, in cases where this ratio is not 1:1, we instinctively seek a specific reason or the cause of that difference. But just why does a 1:1 sex ratio exist among such a large number of animals? As a ratio of almost 1:1 is observed among humans as well, we consider this to be a normal phenomenon. Yet, when we consider the situation in detail, this ratio is definitely not a matter of course. For prosperity of a species, a situation wherein the number of males is considerably less than that of females is perfectly acceptable. This is because the population size in the next generation is constrained by the number of females, whereas males can derive offspring by copulating with several females. Therefore, the number of males is seldom a constraining factor for the number of offspring. The sex-ratio issue cannot be explained in terms of the "prosperity of a species." The fact that the sex ratio is 1:1 is explained in terms of an individual's fitness and not the fitness of the population (species).

To simplify this explanation, let us consider a hypothetical example. Assume that a bird species in which each female lays the same number of eggs once, fledges its young and then dies. Now, consider that a major proportion of the population is female. How many grandchildren will be produced from one female individual? The daughters will produce grandchildren according to their ability. Since the number of females in this population is more than that of males, the sons could potentially mate with several females, and the number of grandchildren produced by each son can be estimated as follows:

Number of grandchildren = Number of grandchildren produced by each female × Number of females that the son copulated with.

Therefore, the number of grandchildren that are produced by each son outnumbers the number of grandchildren produced by each daughter. Females that produce more sons will have more grandchildren. In the population in which a major proportion is female, the descendant of a female that produces more sons will prosper. After several generations, then the number of males in the population will rapidly increase. Conversely, if males outnumber females, the situation will be reversed, and it will be more advantageous to produce more daughters. Females can steadily produce offspring according to their ability, whereas not all males be able to copulate with females; therefore, on an average, the number of grandchildren produced through daughters exceeds the number of grandchildren produced through sons.

When the number of sons and daughters is even, then the average number of grandchildren produced through daughters and sons is equivalent. Under even sex ratio, females will not have the advantage of producing biased sex. Hence, the sex ratio is 1:1[1]. However, note that, although this argument is true for the population sex ratio , it cannot be used to predict whether or not each individual will produce offspring in the ratio of 1:1. Even if the number of females that produce only sons and those that produce only daughters is equal, the fitness of each female will be equal.

This explanation of 1:1 population sex ratio was first proposed by R. A. Fisher[1]. To be exact, he predicted that females will produce offspring at a sex ratio whereby the maternal investment toward sons and daughters will be equal. Moreover, when a difference in the mortality rate between sons and daughters arises while they are investing in their offspring (i.e., under parental care), because the investment toward sons and daughters changes, the sex ratio might be adjusted according to the Fisher's prediction. According to Fisher's predictions, if more sons die, then more sons will be produced, and at the termination of parental care, daughters outnumber sons. However, sexual differences in the mortality rate that arise after maternal investment is completed do not affect the sex-ratio adjustment[2].

Males are abundant among the Rufous Vangas

In the previous chapter, we described that some Rufous Vanga pairs are accompanied by helpers and that these helpers are almost exclusively male. This implies that in the study population, males were clearly more abundant than

Table 7-1 Population sex ratio (rate of males to all individuals) (modified from Asai et al. 2003[8])

Research year	Yearling male	Two-or more year-old male	Female
1994	6	17	19
1995	10	48	42
1996	13	48	43
1997	20	61	52
1998	11	69	50
1999	10	66	44
2000	6	57	33
Total	76	366	283
Sex ratio (including one-year-old)		0.61	
Operational sex ratio		0.56	

In both the sex ratio (including one-year-old) and operational ratio a statistically significant difference from a 1:1 sex ratio was present.

females. Table 7-1 shows the population sex ratio, and it is observed that males outnumber females every year. The sex ratio among sexually mature and potentially fertile individuals is termed the Operational Sex Ratio (OSR). In Chapter 6, it was noted that yearling males of the Rufous Vanga are most probably sexually immature. Although the OSR shown in Table 7-1 excluded yearling males, it is nevertheless male biased.

As mentioned in the previous section, the sex ratio of numerous animals is approximately 1:1. The male-biased sex ratio of the Rufous Vanga is statistically significant, and therefore, we cannot attribute it to chance alone. In cooperatively breeding birds, such as the Rufous Vanga, the sex ratio tends to be biased toward the sex of the helpers. In order to explain this bias, S. T. Emlen et al. proposed a "Repayment Model Hypothesis"[3]. Among several cooperatively breeding birds, the individual that assists the pair (helper) is the pair's offspring produced during the previous breeding season. In the course of the year after it was hatched, it does not disperse but remains within the natal territory and assists the parents. Moreover, the helper's assisting behavior enhances the pair's reproductive success. According to the Repayment Model Hypothesis, in such a situation, a breeding female can enhance the possibility of having helpers in future by biasing the sex ratio of a clutch toward the sex that becomes the helper. The breeding female can thereby improve its own reproductive success.

In keeping with the predictions of this hypothesis, there exist several cooperatively breeding birds, including the Red-cockaded Woodpecker *Picoides borealis*[3] and the Seychelles Warbler *Acrocephalus seychellensis*[4], among which the sex ratio is biased. In the case of the Rufous Vanga, the helper is male, and a male-biased sex ratio is observed; therefore, it appears that the Repayment Model Hypothesis is applicable in this instance. A significant aspect of this hypothesis is that the reproductive success of pairs increases due to the assisting behavior of the helper. However, as noted in the previous chapter, there is no evidence suggesting that the assisting behavior has any significant effect in the case of the Rufous Vanga. If the helper individual does not improve the pair's reproductive success, then the Repayment Model Hypothesis is inappropriate to explain the sex-ratio bias observed in the case of the Rufous Vanga. Ultimately, we must conclude that this species realizes a 1:1 sex ratio as per Fisher's Hypothesis.

Bias in the population sex ratio

If the sex ratio is determined according to the Fisher's Hypothesis, there should be no difference between sons and daughters in the benefit accrued. Consider that the number of grandchildren that its children produce in their lifetime as benefit for the mother. As mentioned earlier, among the Rufous Vangas, males become independent and reproduce after they have remained within the natal territory and assisted for a year or more, whereas females begin reproducing at age one. Therefore, when a male begins breeding, its female siblings have already reproduced at least once. Under such conditions, it is expected that daughters will always be able to produce more grandchildren than will sons. Why then do mothers not produce more daughters? Does the male-biased sex ratio of our study population contradict Fisher's Hypothesis?

Fisher's Hypothesis addresses the sex ratio of offspring at fertilization; thus, rather than comparing the sex ratio in the adult bird population under study, we should examine the sex at fertilization. Moreover, if the investment toward the sons and daughters is extremely differentiated, then this may also be a factor in biasing the sex ratio. The investment (costs of parental care) must also be considered.

We needed to determine the sex ratio at fertilization; however, at present, there exists no method whereby this can be reliably ascertained. However, the sex ratio at hatching is considered an acceptable substitute for this information, and we therefore examined this instead. Even after hatching, the sex of chicks cannot be evaluated until after they have reached a certain level of maturity. Moreover, in order to approximate the sex ratio closest to that at fertilization, the possibility of a biased sex ratio in the nest due to egg and chick mortality must be excluded. Therefore, brood sex ratio should be examined only in nests wherein egg or chick mortality did not occur. However, since brood size often decreased due to predation, the number of nests available as subjects for sexing decreased. As mentioned in Chapter 3, the Rufous Vanga is sexually dimorphic

Table 7-2 Numbers and sex ratio of chicks (ratio of males: to all individuals) (modified from Asai et al. 2003[8])

Research year	Nestlings		Fledglings		
	Male	Female	Male	Female	Sex unknown
1994	4	2	6	4	2
1995	3	4	11	19	1
1996	17	18	18	17	0
1997	19	11	15	10	0
1998	36	32	18	9	0
1999	23	19	14	9	1
2000	14	6	8	5	2
Actual count	116	92	90	73	6
Sex ratio	0.56			0.55	

In sex ratio there is no difference from that of Table 7-1, however, statistically significant bias was not evident.
Numbers of nestlings are derived from nests in which clutch size did not decline, and fledglings indicate the number of whole population, irrespective of any decline in clutch size (refer to text).

in plumage coloration, but we were able to assign sex on the basis of plumage coloration only after birds had reached one year of age. Since some targeted individuals dispersed from the study area soon after fledging, sexing them was impossible. Therefore, we evaluated sex by sampling blood from approximately seven-day-old chicks, extracting DNA from the blood, and examining the genes of the sex chromosomes. With the introduction of this method, nests wherein chicks survived until six days after hatching were added to the research database, and there was a dramatic increase in the amount of viable data. Yet, the possibility of predation of the eggs and chicks of the Rufous Vanga is extremely high, and the number of nests wherein chicks survived until six days after hatching was low. Ultimately, a sufficient number of nests from which we could statistically analyze clutch sex ratio within this population could be obtained only by incorporating all relevant data accumulated over the entire study period.

Table 7-2 shows the sex ratio of the study population at hatching. This consists of data on nests wherein the sex of every chick was determined, and it excludes those nests wherein the number of eggs laid or the number of chicks had decreased by the time we sampled the chicks' blood. This table indicates that male chicks are more numerous than female chicks; however, this difference was not statistically significant. Moreover, even in cases wherein all fledglings in the population, no sex bias was observed. Thereby, it was revealed that the bias in the population sex ratio.

Sexual difference in lifetime reproductive success

One of the most significant aspects of Fisher's Hypothesis concerns the number of grandchildren that will be produced by sons and by daughters. To elucidate this, it is necessary to investigate reproductive success of each individual throughout its lifetime. However, if we attempt to collect such desirable data, then it would be necessary to conduct an extremely long-term research project. Instead, here, we infer this information by averaging the seven years of data, considering the reproductive success of each year as being almost equal.

The accumulated reproductive success up to a certain age = Σ (Survival rate up to each age × Possibility of breeding × Average reproductive success). By examining the manner in which the accumulated reproductive success changes in relation to sex, we are able to infer whether it is more advantageous to bear sons or daughters. The data required to calculate this includes: (1) survival rate, (2) possibility of breeding, and (3) reproductive success. Each of these values is described as follows:

(1) Survival rate is the opposite of mortality rate. First, I will explain the inference of the mortality rate because a consideration of the mortality rate is intuitively easier to understand. In most cases, it is not possible to determine when an individual died. Even in cases wherein we lost track of a leg-banded individual, we were unable to establish whether this was due to mortality or due to the individual having dispersed and taken up residence outside the study area. If we can estimate the percentage of disappeared individuals that can be attributed to mortality, it will be possible to explain the average mortality rate in the research population. A large proportion of the Rufous Vangas in the research population had been individually identified (75-94%); therefore, the number of individuals that immigrated from outside the study area could be specified because these had not been leg banded. Unless the conditions within the research area and the location from which the bird came differ greatly, it is considered that there is no difference in an individual's immigration/emigration situation, reproductive success, survival rate, etc. In fact, the research area and its surrounds are almost completely forested, as shown in Plate I-4. Therefore, if we hypothesize most simply that the number of individuals that emigrated from the study area and the number of individuals that immigrated into the research area is equal, then

"The number of disappeared individuals − The number of immigrated individuals = The number of deceased individuals."

We used this formula to calculate the annual mortality rate.

Table 7-3 Estimated mortality rate averaging seven years (modified from Asai et al 2003[8])

Male (a)	Female (b)	Breeding male (c)	Helper male (d)	Fledgling male (e)	Fledgling female (f)
0.16	0.14	0.21	0.13	0.26	0.61

The letters correspond to the explanation in the text.

Mortality rate was calculated for each year, and the mean value was used as inference of the accumulated reproductive success. Mortality rate was calculated separately for six categories of individuals: (a) males at age of one year or more, (b) females at age of one year or more, (c) breeding males, (d) helper males, (e) fledged males, and (f) fledged females. (a) We calculated the mortality rate of males at age of one year or more literally, without discriminating among males. However, since it was expected that mortality rate differs between breeding males and helper males, we divided these further into categories (c) and (d) and calculated mortality rate separately. As shown in Figures 3-7 and 3-8, once breeding males acquire a territory, they do not disperse; hence, we considered all immigrated individuals to have been helper males originally. Therefore, in (c), the number of the disappeared individuals are the number of deceased individuals. Categories (e) and (f) present the values that the individual number of yearlings is divided by the number of chicks fledged the previous year.

From Table 7-3, it is clear that the mortality rate of fledged females is particularly high, which implies that a sex-related difference in the mortality rate occurs between the time of fledging and the yearling stage. In general, when young birds move away from the natal territory shortly after fledging, their mortality rate is high. Females disperse from the natal territory immediately after fledging and begin breeding when yearlings. In contrast, the mortality rate of helper males is the lowest among all the examined categories, which implies that males increase their survival rate by remaining in the parent's territory as helpers.

(2) The possibility that a male is a breeder was derived from the number of breeders in a year divided by the potential number of breeders. Yearling males are not considered sexually mature (Chapter 6); therefore, individuals in this category were not included in the number of potential breeders. In the case of males, the possibility of them being breeders was calculated separately for individuals between one and two years of age and those over two years.

The accumulated reproductive success of males was calculated in two ways:

Table 7-4 Reproductive success in the studied population

Research year	Number of fledglings	Number of breeding pairs	Number of chicks per breeding pair
1994	12	10	1.20
1995	31	24	1.29
1996	35	27	1.30
1997	25	44	0.57
1998	24	43	0.56
1999	24	33	0.73
2000	15	27	0.56
Average	23.71	29.71	0.89

The pairs of which reproductive success/failure are uncertain were excluded.

(i) by randomly deciding which male was able to breed and (ii) by dividing a male's lifetime into two periods, helping and breeding, which more closely approximate the actual life history of males. It is necessary to recalculate reproductive success separated by breeding age. In the former case, whether the individual is a helper or a breeding male was disregarded, and the calculation was performed considering all males as having an equal possibility of breeding. When the latter was calculated, the possibility of a male being a breeder was calculated as zero during the helper period, and thereafter, the possibility was calculated as 1 during the breeding period.

Based on this protocol, the survival rate was calculated in two ways. In the case wherein determination of whether an individual was a breeder or not was randomly decided, calculations were based on (a) the mortality rate of males at age of one year or more. Alternatively, on the basis of (c) and (d), survival rate was calculated for two periods depending on an individual's life history. Then, using the respective mortality rates, the accumulated reproductive success was calculated.

(3) In terms of the reproductive success, Table 7-4 indicates the number of chicks fledged by each pair during every breeding season. The mean value of each year is used to calculate the accumulated reproductive success.

As a result, the plot shown in Figure 7-1 was obtained. This figure shows that males beginning breeding at age two would experience a higher lifetime reproductive success than the average female. Moreover, if an individual male that begins breeding at age four, has a lifespan of nine years and repeats breeding, it would be able to obtain a lifetime reproductive success rate comparable to that of average females. However, if males begin breeding at age four and continue only until less than nine years of age, or if they begin breeding

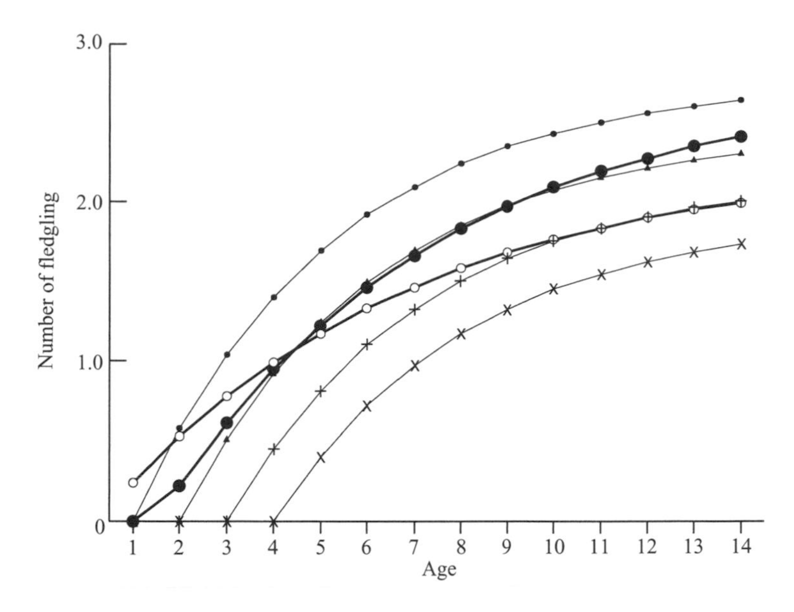

Fig. 7-1 Transition of accumulated reproductive success

○ : Female

The probability that whether the individual can become a breeding individual or not was determined randomly.

● : Male

Concerning the life history from a helper to a breeding male.

• : Male which breeds from two-year-old

▲ : Male which breeds from three-year-old

+ : Male which breeds from four-year-old

× : Male which breeds from five-year-old

after age four, the lifetime reproductive success rate would be lower than that of the average female.

Males that begin breeding at age three would be able to acquire almost the same lifetime reproductive success as an average female. At what age do males actually begin breeding? The age at which onset of breeding occurred was known for 24 male individuals, among which 15 began breeding at age two. The mean age at which onset of breeding occurred was 2.7 years. Moreover, in the year 2000, the final year of our research, there was one six-year-old male individual that was still a helper. The research period was limited to seven years; therefore, our estimates of the age at which the onset of breeding occurs are unavoidably biased toward the younger extremes. Perhaps the mean age at which males begin breeding is approximately three years. If so, this implies that the average lifetime reproductive success of males and females is almost equal. Moreover, the accumulated reproductive success calculated for the group wherein the age at which the helper changes into a breeding male was disregarded form almost the same curve as that of the group of males wherein onset of breeding occurred at age three. Therefore, it was implied that the average lifetime reproductive success of males and females is almost equal irrespective of whether the life history of males is considered or not.

Because the accumulated reproductive success was calculated on the estimates after fledging, the result of Figure 7-1 is applied to individuals that fledge successfully. The eventual equalization in the lifetime reproductive success of males and females results from a difference in the survival rates between the male and female after fledging. A large proportion of female individuals die after fledging before they have the opportunity to breed. The expected fitness of daughters at fledging is decreased due to a high subsequent mortality rate. In contrast, since the male survival rate is higher than that of females, on an average, the expected fitness of sons at fledging decreases only slightly, even if the onset of breeding is late.

Sex-related differences in parental care costs

As mentioned in the previous section, at fledging, there is no sex difference in the expected future reproductive success. This implies that after a major proportion of parental care has been administered, there exists no difference between sons and daughters with regard to the number of grandchildren that the mother expects. However, if there is a difference between sons and daughters in parental care cost, then there should be a sex bias toward the less costly sex.

The relationship between the number of female chicks and the frequency at which the nest is provisioned (Fig 7-2) indicates that even if the number of female chicks increased, there would be no trend of increase or decrease in the frequency of provisioning food. Although not shown in the figure, no correlation was observed even when the provisioning frequency by the breeding female was examined separately. However, the provisioning frequency data was collected only during the latter half of the parental-care period. Further, if change actually occurs not in terms of the provisioning frequency, but in the volume of food provided to the chicks at each visit, then it would not be possible to detect any change in the investment toward the chicks. Alternatively, each chick's daily growth and the size before fledging was calculated (Figs. 7-3 and 7-4) on the basis of data that was obtained after the chicks were more than six days old. According to these figures, the only statistically significant difference between these groups was observed in the growth of wing length. Accordingly, the presence of a sex-related difference in parental cost by the time of fledging

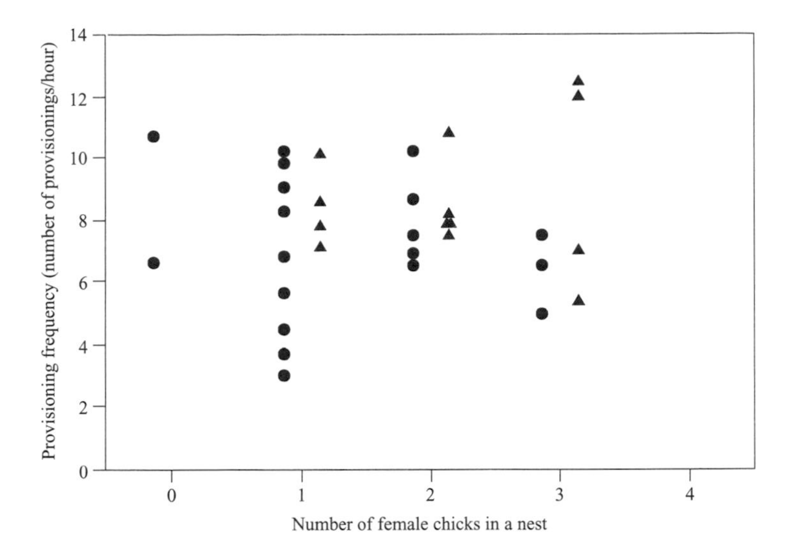

Fig. 7-2 Sex ratio and provisioning frequency in the nest
● : Nest of three chicks
▲ : Nest of four chicks
No correlation was evident between the parameters.

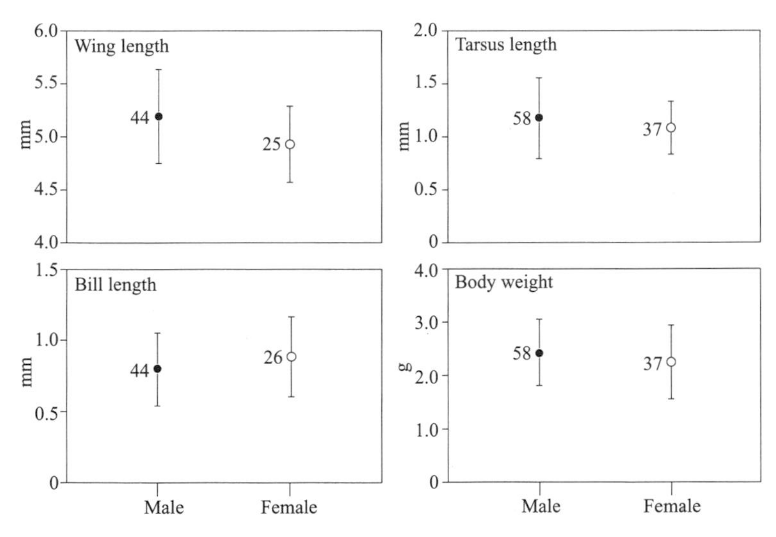

Fig. 7-3 Daily growth increment of chicks
Numeric values in the figure indicate the number of samples, and bar indicates standard deviation. Except for wing length, no statistically significant difference between male and female was evident.

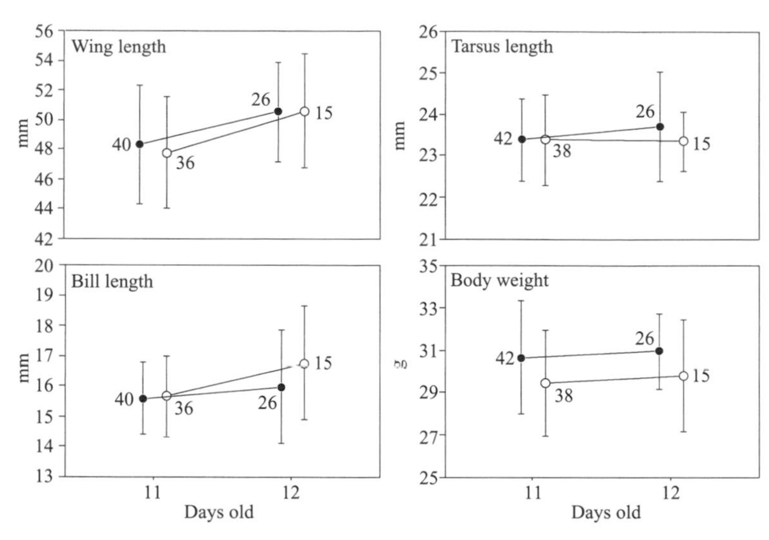

Fig. 7-4 Body size of chicks
11-day-old chicks and 12-day-old chicks are different individuals. The numeric values in the figure indicate the number of samples, and bar indicates standard deviation. No statistically significant difference was evident between males and females in either 11-day-old chicks or 12-day-old chicks.

seems improbable, and it is implied that for breeding females, there is no difference in the investment between sons and daughters.

I will now summarize the results obtained thus far. Considering that among the Rufous Vangas, females can breed earlier than males, it would appear more beneficial for a breeding female to produce more daughters than sons. However, the sex ratio in the studied population was strongly male-biased, which seemingly contradicts this supposition. Investigations revealed that the sex ratio at hatching was 1:1 and that the bias in the population was the consequence of a sex-related difference in the mortality rate. If calculated from the point of fledging, the average lifetime reproductive success of males is never lower than that of females. When they fledge, a breeding female can expect the average lifetime reproductive success of its sons and daughters to be equal; further, since the cost of parental care is not different between sons and daughters, the breeding female does not bias the sex ratio of a clutch toward either sex. Such findings agree with the predictions of Fisher.

The case wherein the mother biases the sex ratio toward sons or daughters

Thus far, the above explanation might have given readers the impression that each female Rufous Vanga lays eggs at the ratio of 1:1 (sons:daughters). However, this discussion focused only on the overall population, and the situation wherein each individual female biases the sex of its offspring has not been considered. The sex ratio at hatching signified the number of birds present in that year's cohort of the entire population. The accumulated reproductive success was also calculated from the mean value of the entire population. The explanation in Section 7.1 states that the sex ratio becomes 1:1 because, on an average, when males outnumbers females in the population, it is more advantageous to bias the sex of offspring toward females and vice versa. Therefore, a stable 1:1 sex ratio can be applied to only the outcome of averaging the entire population. Consequently, it is not considered whether all females in the population produce chicks at a 1:1 sex ratio or whether females that produce only sons and females that produce only daughters exist at the ratio of 1:1. Could it be possible that females determine the sex of their chicks and that the sex ratio in the population at hatching is adjusted to 1:1 by means of a more complex pattern?

R. L. Trivers and D. E. Willard demonstrated that it is more advantageous for the mother to bias the sex ratio of its offspring depending on its situation[5]. For instance, among polygamous animals, males exhibit large individual differences in fitness, and those in good physical condition (such as having large body size, etc.) can obtain more females. Moreover, as observed in the case of mammals, when the mother invests in its children over a long period, the physical condition of the mother is believed to greatly influence the physical condition of the offspring. It is predicted that when the mother's physical condition is good, the child will be nursed well. Thus, it would be more advantageous to produce sons since a male that is in better physical condition has a high possibility of copulating with more females. As a result, such a male is expected to produce more offspring. Conversely, for a female in poor physical condition, it is more advantageous to give birth to daughters since it is expected that females can definitely produce offspring. In competition with other males over the acquisition of a female, males in poor physical condition might be defeated by individuals in superior condition, and they may ultimately be unable to produce any offspring whatsoever.

On the basis of the theory of Trivers and Willard, it has been elucidated that several animals bias the sex ratio of their offspring depending on the situation. For example, the Great Reed Warbler occurring in Japan is a polygamous bird;

the female adjusts the sex in a clutch, depending on its situation[6]. Among the Great Reed Warblers, several females have nests within the territory of one male; however, the male carries food only to the nest of the female with which it first copulated (first female). Accordingly, only the nest of the first female has an abundance of food, and the offspring in the nest are well nurtured. Moreover, it is considered that in the case of this species, the better the physical condition of the male, the greater the number of females it can copulate with. The first female will produce more sons because they are expected to have the better physical condition. In contrast, the second female produces more daughters.

Moreover, it is believed that in addition to the physical condition, there is a genetic element involved. Males that copulate with several females probably have genetically advantageous characters, and females that copulate with such males can expect their sons to inherit these advantageous genes. Consequently, when the mating partner copulates with several females, the females will bias the sex ratio of the clutch toward males.

Among other bird species, it has been recognized that females adjust the sex ratio of a clutch depending on the situation. However, the situations that induce adjustment of sex ratio are diverse. Sex-ratio bias occurs both when the fitness of males exhibits large variations as compared to that of females, and when individual differences among males with regard to fitness are influenced by the mother's situation.

The Rufous Vanga female can bias its offspring toward sons or daughters

Does the Rufous Vanga bias the sex ratio of its offspring? As shown in Figure 7-1, it is inferred that the lifetime reproductive success of Rufous Vanga males exhibits large variation as compared to that of females, depending on the age at which breeding begins. For a male that assists at the parents' territory, it is preferable to become independent and begin breeding immediately after finding a mate. However, due to the presence of several other males also awaiting mates, becoming independent could be different. If sons are unable to early locate a mate, it is more advantageous for the mother to produce daughters. This is because the lifetime reproductive success of an average female is higher than that of sons that cannot find partners by age three or four. Conversely, in a situation wherein female mates are easily available, it is more advantageous for the mother to produce sons. This is because, as shown in Figure 7-1, if sons can begin breeding at age two, the mother can expect them to obtain a higher lifetime reproductive success than that of average females. What are the conditions under which sons can find partners easily?

Table 7-5 Number of nests which showed a biased sex ratio
In the category of With helper, the bias from expectant was not significant, however, in the category of Without helper, it was significantly biased.

Sex allocation	With helper	Without helper
male > female	7	19
male = female	9	6
male < female	11	3

As mentioned earlier, almost all helpers are males without a mating partner. Therefore, for a female just producing eggs, the presence or absence of helpers is considered to be the simplest criterion to assess whether or not its sons would find partners easily.

We recorded 25 cases of male helpers that became independent and began breeding (the number is different from that provided in Table 6-3 due to the availability of a larger database). Among them, one individual took over a territory from a breeding male; however, in all other cases, these males established territories in the neighboring territory or one territory distant. In 10 out of 11 cases, males that immigrated from outside the study area established territories at the periphery of that area. Consequently, it is considered that when a helper becomes independent, it moves only a limited distance from its natal territory. Therefore, its competitor over a mate is other nearby helpers, and the nearest helper is, most often, a helper in the same territory. Moreover, 12 out of 25 individuals became independent from territories that did not have any other helpers. Further, 10 individuals were the oldest helpers in their respective territories when they became independent from territories wherein other helpers were present. Only two younger individuals could become independent. The age relationship of the remaining individual with other helpers was unknown. Thus, helpers can become independent when they are the only helper or the eldest of the helpers present. Therefore, when a helper is already present, it is highly probable that the youngest males are at a disadvantage in terms of male-male competition over a mate.

Table 7-5 shows the sex allocation in nests categorized by whether they were accompanied by a helper or not. The sex ratio is biased toward females in nests where the breeding pair was accompanied by a helper. A statistical test reveals that it is unlikely that this bias occurred by chance. The nests shown in Table 7-5 do not include the nests wherein predation caused a decrease in eggs and chicks; that is, the possibility of a sex bias resulting after the eggs were laid is ruled out. Thus, the results indicate that this bias is present at egg laying.

How does the female bias the sex of its offspring? Several possible stages exist in the adjustment of offspring sex, one of which is the manipulation of the combination of chromosomes during fertilization. Among birds, males have the same sex chromosome type: ZZ; therefore, gametes produced by males always possess Z chromosomes. On the other hand, females possess ZW chromosome type, and there are two types of gametes with Z chromosomes and another with W chromosomes. Therefore, offspring sex is determined by the chromosome that the gamete deriving from the female has. If the female has the ability to choose the ovum for fertilization on the basis of the knowledge of the chromosome composition of the gamete (ovum), it will be able to adjust the sex ratio during fertilization. However, the precise mechanism of sex-ratio adjustment is not known. Alternatively, they may be able to adjust the sex ratio by judging the sex after fertilization and by inhibiting the development of one particular sex.

Biased sex ratio at hatching and delayed dispersal

The story thus far may be summarized as follows:

The population sex ratio of the Rufous Vanga is male-biased. The Rufous Vanga is monogamous, and surplus males become helpers. Each helper (as well as the breeding male) competes for mates. Since the oldest helper male has an advantage over younger helpers in mate competition, a newly hatched male may opt to delay the onset of breeding if other helpers exist around it. When a female lays eggs, it expects its progeny to produce as many offspring as possible. Therefore, in a situation where no helpers exist, it produces more sons, which have high survival rate, and in a situation where helpers exist, it produces more daughters, which can breed from age one. When sons cannot begin breeding independently and they delay their dispersal, the mother gains a helper. For the mother, the optimal situation would be to produce sons whenever the son can begin breeding at age two and to produce daughters otherwise. However, a female that biases its offspring largely toward sons faces the risk of the sons' onset of breeding being delayed because of competition over mates. Accordingly, it would be preferable to bias in a subtler manner. The timing when a helper can become a breeder will be affected by the shift in the population as a whole. The presence or absence of helpers should serve as a parameter of the shift in the entire population. Ultimately, whether or not a helper is present will be a good index for an individual female's sex-ratio adjustment. Because an individual female's sex-ratio adjustment is closely related to the population sex ratio, even if an individual female allocates the sex of its offspring based upon the presence or absence of helpers, the population sex ratio at hatching is maintained at 1:1.

Most commonly, the onset of breeding among males occurs at age two; perhaps, this may be because mothers adjusted the sex-ratio in their clutches successfully. However, it is not always successful. Hence, helpers over two years of age are always present within the population. Because the population sex ratio at hatching is 1:1, males will be in surplus due to the ecological trait of a sex difference in the mortality rate of fledging chicks. Yearling males are considered to be sexually immature, and in this case, they inevitably do not participate in breeding. Therefore, it is not accurate to state that they cannot breed because of ecological restraints. However, from a general perspective, small birds, such as the Rufous Vanga, can usually begin breeding at age one. Therefore, it may be more accurate to state that since an ecological constraint prevents them from breeding at age one, sexual maturity is delayed. Given that the younger the age at which the onset of breeding occurs, the greater the individual's lifetime reproductive success, it is considered that males are refrained from breeding because of ecological factors. The topic of the sex ratio adjustment has been considered from the perspective of the mother, not the offspring. It has already been noted that on an average, when sons are able to breed from age three, their lifetime reproductive success will be equal to that of daughters. However, from the perspective of the sons, the optimal strategy is to begin breeding as soon as they attain sexual maturity. However, since the optimal sex ratio from the mother's perspective has been achieved, this strategy is not realized. In this regard, there exists a conflict of interest between mother and son. Moreover, in such a scenario, the mother possesses greater advantage and is dominant over the son.

It is considered that for males with no option but to delay their breeding, remaining within the parental territory to increase their viability is the optimal tactic to increase fitness. The mortality rate of helpers was lower than that of any other category of individual (Table 7-3). Moreover, helpers that remained in their natal territory exhibited helping behavior characteristic of the social system termed cooperative breeding. In this context, it is found that the issue of sex-ratio adjustment is closely related to the maintenance of the cooperative-breeding social system.

Do males that are restrained from breeding accept the current situation passively? As shown in Figure 7-1, it is considered that by delaying the onset of breeding by a year, males decrease their lifetime reproductive success. To compensate for this cost, a helper may not only undertake helping but may also endeavor to copulate with the breeding female in order to produce offspring. Helpers remain at their parental territory; therefore, in major cases, the breeding female is their own mother. In such a case, the copulation is incestuous, and generally speaking, disadvantageous, because the resulting combination of genes is deleterious to the offspring. Therefore, such cases are usually avoided. Yet, in the case of the Rufous Vanga, the pair female occasionally changes; therefore, it is possible that the breeding female residing in the territory is unrelated to the helper. In such a situation, the helper could possibly obtain its own offspring, and there is no reason for a sexually mature helper to avoid copulation (Chapter 6). The Stripe-backed Wren *Campylorhynchus nuchalis*, found in Venezuela in South America, has long been studied as a cooperative breeder, and among this species, helpers sometimes obtain their own offspring[7]. Although we have not yet detected it, a similar situation may also exist among the Rufous Vangas.

References

1 Fisher, R. A. (1930) The Genetical Theory of Natural Selection. Clarendon Press, Oxford.
2 Charnov, E. L. (1982) The Theory of Sex Allocation. Princeton University Press, Princeton.
3 Emlen S. T., Emlen, J. M. and Levin, S. A. (1986) Sex-ratio selection in species with helpers-at-the-nest. American Naturalist 127: 1-8.
4 Komdeur, J., Daan, S., Tinbergen, J. and Mateman, C. (1997) Extreme adaptive modification in sex ratio of the Seychelles warbler's eggs. Nature 385: 522-525.
5 Trivers, R. L. and Willard, D. E. (1973) Natural selection of paternal ability to vary the sex ratio of offspring. Science 179: 90-92.
6 Nishiumi, I. (1998) Brood sex ratio is dependent on female mating status in polygynous great reed warblers. Behavioral Ecology and Sociobiology 44: 9-14.
7 Rabenold, P. P., Rabenold, K. N., Piper, W. H., Haydock, J. and Zack, S. W. (1990) Shared paternity revealed by genetic analysis in cooperatively breeding tropical wrens. Nature 348: 538-540.
8 Asai, S., Yamagishi, S. and Eguchi, K. (2003) Mortality of fledgling females causes male bias in the sex ratio of the Rufous Vanga (*Schetba rufa*) in Madagascar. Auk 120 : 700-705.

Tracking the route taken by Rufous Vangas

The bird most closely related to the Rufous Vanga (right) is the Helmet Vanga (left).

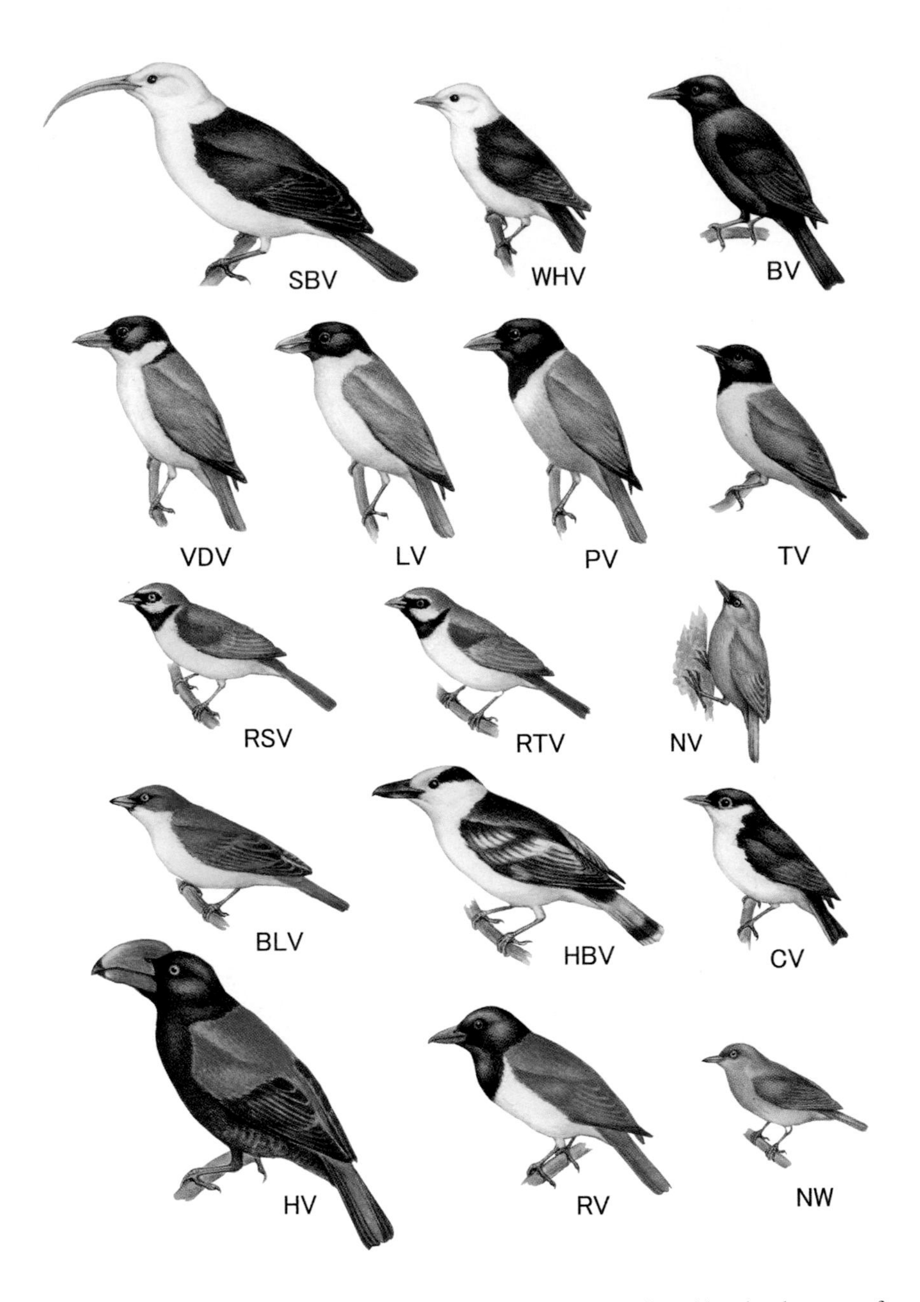

Plate III-1 Birds of Family Vangidae. Refer to Table 8-1 regarding abbreviated names of species (Pictures redrawn by K.Kanao from Langrand, 1990)

Tracking the route taken by Rufous Vangas

Satoshi Yamagishi and Masanao Honda

Rufous vanga and its relatives

The main subject of this book, Rufous vanga *Schetba rufa*, is endemic to Madagascar and belongs to the family Vangidae. *Guide to the Birds of Madagascar* (Langrand 1990)[1] is currently the most popular source of information on these birds. According to this book, the family Vangidae on Madagascar consists of 14 species, of which only the Blue Vanga *Leptopterus madagascarinus* also occurs on the Comoros Islands. After publication of this volume, *Hypositta perdita*, which is similar to the Nuthatch Vanga *Hypositta corallirostris*, was described from museum specimens, though, this species is now widely regarded as being extinct[2]. Moreover, it was recently described that the southwestern population of Red-tailed Vanga *Calicalicus madagascariensis* is a valid species, Red-shouldered Vanga *Calicalicus rufocarpalis*[3], as it was the Comoros Islands' population of Blue Vanga as *Cyanolanius comorensis*[4] (Table 8-1). However, the last of these is not referred to in this study because there is no justification for its classification as a species. Therefore, it is reasonable to currently regard the Vangidae as consisting of 15 species, adding only the Red-shouldered Vanga to the above mentioned 14 species.

When we trace the history of the classification of the family Vangidae, we find that until the French ornithologist J. Delacour recognized "Vangidés" as being endemic to Madagascar[9], most species within this family had been assigned to the family Laniidae (shrikes) (Table 8-2). This is probably why the Vangidae was, until recently, considered to be closely related to the Laniidae. However, the family name "Vangidés", coined by Delacour, was inappropriate because it was a French word. It was the American ornithologist, A. L. Rand, who renamed the family "Vangidae" in accordance with the rules of the

Table 8-1 15 species of family Vangidae. Generic names follow Peters' Checklist of the Birds of the World. Abbreviated names were taken from English names, and they correspond to respective names of Plate III-1. If our study achieves recognition, the family Vangidae will consist of 19 species, including four species of the genus *Newtonia*. The Common Newtonia is included owing to its being presumed to be a member of the family Vangidae this time.

Name of species	English names	Abbreviated names	Body length (cm)
Calicalicus madagascariensis	Red-tailied-Vanga	RTV	13.5 ~ 14
C. rufocarpalis	Red-shoulderd Vanga	RSV	15
Schetba rufa	Rufous Vanga	RV	20
Vanga curvirostris	Hook-billed Vanga	HBV	25 ~ 29
Xenopirostris xenopirostris	Lafresnaye's Vanga	LV	24
X. damii	Van Dam's Vanga	VDV	23
X. polleni	Pollen's Vanga	PV	23.5
Falculea palliata	Sickle-billed Vanga	SBV	32
Leptopterus viridis	White-headed Vanga	WHV	20
L. chabert	Chabert's Vanga	CV	14
L. madagascarinus	Blue Vanga	BLV	16
Oriolia bernieri	Bernier's Vanga	BV	23
Euryceros prevostii	Helmet Vanga	HV	28 ~ 30.5
Hypositta corallirostris	Nuthatch Vanga	NV	13 ~ 14
Tylas eduardi	Tylas Vanga	TV	20
Newtonia brunneicauda	Common Newtonia	NW	12

International Code of Zoological Nomenclature[13]. W. J. Bock[11], however, claimed the Vangidae was first described by W. Swainson in a paper published in the first half of the 1800s[12]. However, having been unable to locate a copy of this text, it has been impossible to verify Bock's claim.

The family Vangidae, established by Rand, is derived from the generic name of *Vanga curvirostris*, which is the scientific name of Hook-billed Vanga (Fig. 8-1). The specific epithet "*curvirostris*" is Latin for "curved bill," "vanga" is the Malagasy word for "spot," and "vanga vanga" means "black and white spot." Thus, the nomenclature signifies "black and white bird with a curved bill." This explanation of the source of the original family name clarifies the fact that the name was not intended to resonate with the name "Laniidae".

Although Rand established the family Vangidae, his account is quiet different from the current classification, which includes several species formerly assigned into other families. In fact, Rand (1936) assigned *Tylas* to the Pycnonotidae, while recognizing the Eurycerotidae and the Hypositidae monotypic with *Euryceros* and *Hypositta*, respectively (Table 8-2). It was the French ornithologist, D. Dorst, who first classified the Vangidae as consisting of the 14 species currently recognized[13]. He argued for the morphological similarities among the members of the family, such as the shape of the skull, jaw muscle attachment, chick's pteryla, scale pattern of the leg, and lumped *Hypositta* and *Tylas* into the Vangidae. However, he neither noted which species had been

Table 8-2 History of classification within the family Vangidae. The classification in this family was highly confused. Dash denotes Vangidae.

Name of species	Catalogue of the British Museum	Delacour[9] (1932)	Rand[10] (1936)	Peters'[5,6,7,8] checklist	Dorst[13] (1960)
Red-tailed-Vanga (RTV)	Laniidae	–	–	–	–
Rufous Vanga (RV)	Laniidae	–	–	–	–
Hook-billed Vanga (HBV)	Laniidae	–	–	–	–
Lafresnaye's Vanga (LV)	Laniidae	–	–	–	–
Van Dam's Vanga (VDV)	Laniidae	–	–	–	–
Pollen's Vanga (PV)	Laniidae	–	–	–	–
Sickle-billed Vanga (SBV)	Corvidae	–	–	–	–
White-headed Vanga (WHV)	Laniidae	–	–	–	–
Chabert's Vanga (CV)	Prionopidae	–	–	–	–
Blue Vanga (BLV)	Laniidae	–	–	–	–
Bernier's Vanga (BV)	Laniidae	–	–	–	–
Helmet Vanga (HV)	Prionopidae	–	Eurycerotidae	–	–
Nuthatch Vanga (NV)	Sittidae	Sittidae	Hyposittidae	Paridae	–
Tylas Vanga (TV)	Pycnonotidae	Pycnonotidae	Pycnonotidae	Pycnonotidae	–

compared nor used any other suboscine family as a point of comparison. Therefore, it is insufficient to conclude that the Vangidae is a monophyly–a taxonomic group within which all descendants originated from a single ancestor. Even recently, some scientists have assigned *Tylas* either to the Pycnonotidae (bulbuls)[14] or to the Oriolidae (Old World orioles)[15].

The high degree of morphological diversity, as evident from an examination of Plate III-1, leads to taxonomic confusion even today. The body length of vangas ranges from a minimum of 13 cm, in the case of Nuthatch Vanga, to a maximum of 32 cm, in the case of the Sickle-billed Vanga—a size variation of almost 250%. Plumage coloration is also diverse, with white, black, blue,

Fig. 8-1 A Hook-billed Vanga *Vanga curvirostris*, visiting the nest. This bird has a black and white two-tone color plumage, and the family name Vangidae derived from this bird.

auburn, and ash gray being observed among the various species. Often, when two species forage in a similar manner, their beak shapes evolve toward a similar form although they belong to different lineages. This phenomenon is termed convergence. It would be erroneous to conclude that these two species belong to the same lineage on the basis of the similarity in bill shape that resulted from convergence. Conversely, morphological differences often evolve between species that possess a common origin but are placed in different environments. This phenomenon is termed radiation. In this instance, it would be incorrect to conclude that these two species belong to different lineages on the basis of morphological differences that resulted from radiation.

How can this dilemma be resolved?

To resolve this dilemma, it is necessary to use other information in addition to morphological data. At present, the most efficient approach is molecular phylogeny. In this analysis we compare the different birds' DNA sequences, which consist of four bases: adenine (A), cytosine (C), guanine (G) and thymine (T). We calculate the degree of affinity from similarities in base configurations,

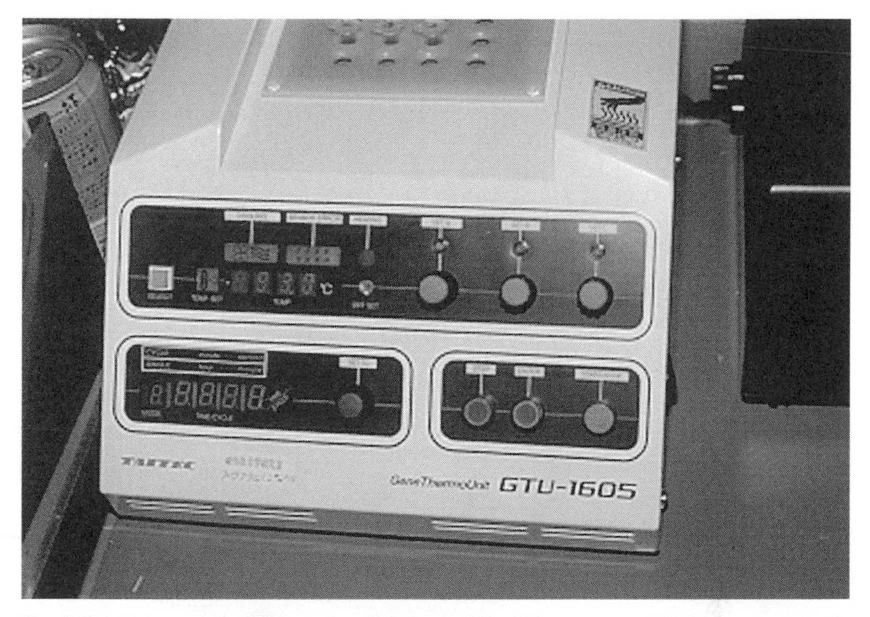

Fig. 8-2 A Polymerase-chain-reaction (PCR) amplifier. The amount of DNA is automatically increased through this device.

and we use this information to produce a dendrogram. This process is based on the development of two methods in molecular biology. One of these is the polymerase chain reaction (PCR) method, which amplifies specific fragments of tiny amounts of DNA. Using this technique, we can obtain large quantities of DNA even from a small sample of tissue or blood without damaging the specimen (Fig. 8-2). The other development that has proven useful is that of auto-sequencer machines, which automatically read DNA sequences (Fig. 8-3). The use of these devices has led to rapid advances in this field of research.

Using this method, we attempted to compare the sequence data from cytochrome *b* gene of mitochondrial DNA, were unable to obtain a satisfactory result[16]. At around the same time, a friend and colleague, the American scientist T. S. Schulenberg, also attempted the phylogenetic analysis of the Vangidae using the cytochrome *b* gene. In his dissertation, he notes that he could not obtain evidence that the family Vangidae is monophyletic[17]. However, his analyses using cytochrome *b* gene suffered problems of saturation of base substitution. We therefore consider that both these analyses using the cytochrome *b* gene were fraught with problems that precluded them from clarifying the phylogenetic relationships within the family.

Consequently, we analyzed approximately 880 base positions of mitochondrial DNA sequences of 12S and 16S ribosomal RNA (rRNA) genes.

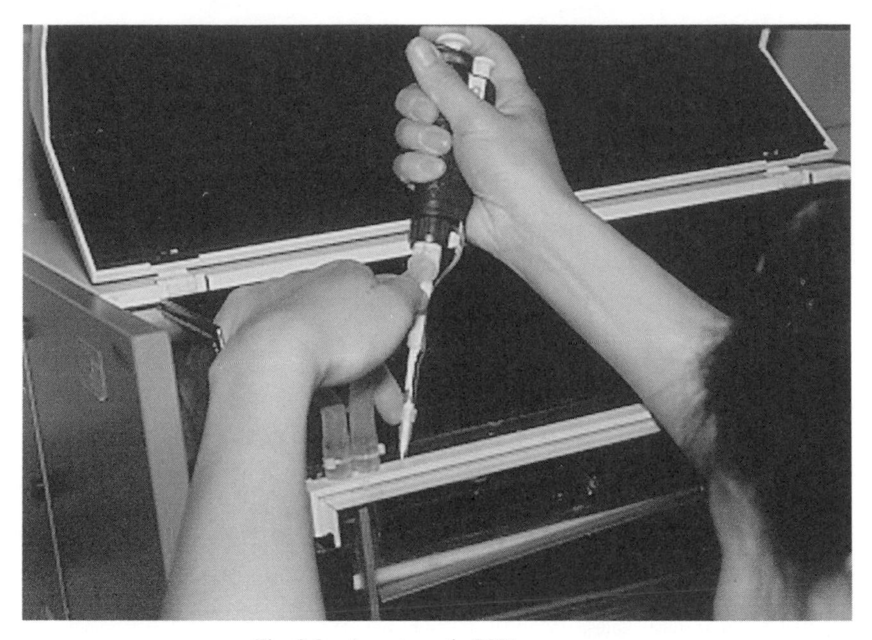

Fig. 8-3 An automatic DNA sequencer.

Although we failed to collect blood samples of Lafresnaye's Vanga *Xenopirostris xenopirostris* and Pollen's Vanga *Xenopirostris polleni* as well as Red-shouldered Vanga *Calicalicus rufocarpalis*, we were able to use the Van Dam's Vanga *Xenopirostris damii* and Red-tailed Vanga *Calicalicus madagascariensis* as the representatives of these two genera. Thus, our analysis included 12 of the 15 existing species, representing all genera within the family. Furthermore, to elucidate the ancestor of the vangas, we incorporated into the analyses all those oscine families, for which tissues were available to us, including *Laniarius ferrugineus* and *Newtonia brunneicauda*. Through the sequencer method (Fig. 8-4), we obtained the sequence data shown in Table 8-3. A careful examination of the table reveals that occasionally certain bases are replaced by bases of a different type. This is termed mutation. In mitochondrial DNA, this substitution occurs at a far greater frequency than in nuclear DNA. Therefore, based on the nucleotide substitution, we can investigate the degree of genetic relationships among closely related species.

Using a phylogenetic estimation method, termed neighbor-joining (NJ) analysis, we arrived at a conclusion that clarified the hitherto confused phylogenetic relationships of the Vangidae[18]. For example, Figure 8-5 strongly suggests that the Vangidae is monophyletic and that Tylas Vanga, which was often classified into the Pycnonotidae (bulbuls), and certain species that were

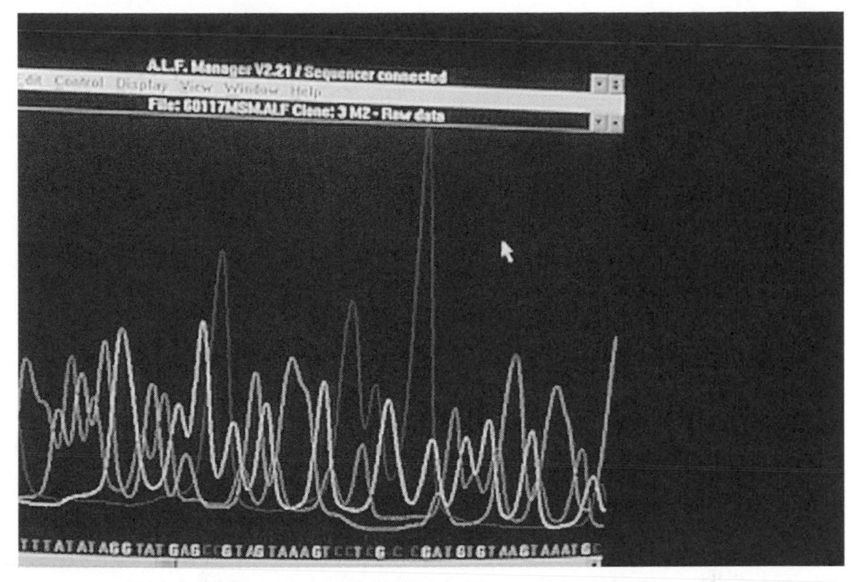

Fig. 8-4 DNA sequencer automatically reads out the sequence of four kinds of bases.

Table 8-3 Sixty base pairs out of 1437 base pairs of 12SrRNA and 16SrRNA are shown. Underlines show occurrences of mutation. Refer to Plate III-1 and Table 8-1 for abbreviated names of species.

BLV	ACAAACGCTTAAAACTCTAAGGACTTGGCGGTGTTCCAAACCCACCTAGAGGAGCCTGTT
BV	ACAAACGCTTAAAACCCTAAGGACTTGGCGGTGCCCCAAACCCACCTAGAGGAGCCTGTT
CV	ACAAACGCTTAAAACTCTAATGTCTTGGCGGTGCCCCAAACCCACCTAGAGGAGCCTGTT
HBV	ACAAACGCTTAAAACTCTAAGGACTTGGCGGTGCCCCAAACCCACCTAGAGGAGCCTGTT
HV	ACAAACGCTTGAAACTCTAAGGACTTGGCGGTGTTCCAAACCCACCTAGAGGAGCCTGTT
NV	ACAAACGCTTAAAACCCTAAGGACTTGGCGGTGTCCCAAACCCACCTAGAGGAGCCTGTT
RTV	ACAAACGCTTAAAACTCTAAGGACTTGGCGGTGCCCCAAACCCACCTAGAGGAGCCTGTT
RV	ACAAACGCTTGAAACTCTAAGGACTTGGCGGTGCTCCAAACCCACCTAGAGGAGCCTGTT
SBV	ACAAACGCTTAAAACCCTAAGGACTTGGCGGTGCCCCAAACCCACCTAGAGGAGCCTGTT
TV	ACAAACGCTTAAAACTCTAAGGACTTGGCGGTGCCCCAAACCCACCTAGAGGAGCCTGTT
WHV	ACAAACGCTTAAAACCCTAAGGACTTGGCGGTGCCCCAAACCCACCTAGAGGAGCCTGTT
VDV	ACAAACGCTTAAAACCCTAAGGACTTGGCGGTGCCCCAAACCCACCTAGAGGAGCCTGTT
NW	ACAAACGCTTAAAACTCTAAGGACTTGGCGGTGCTCCAAACCCACCTAGAGGAGCCTGTT

occasionally placed in other families are members of the family Vangidae. Furthermore, our results indicated that *Newtonia*, a genus endemic to Madagascar that has never before been assigned to the Vangidae, is a member of this family. This is especially surprising because this genus has hitherto been assigned to the Sylviidae (warblers) or to the Muscicapidae (flycatchers). To confirm the NJ dendrogram, we reexamined the relationships using two different methods, maximum likelihood (ML) and maximum parsimony (MP) analyses, and obtained the same result, thereby confirming *Newtonia* to be a member of the family Vangidae.

Tracing the evolution of the family Vangidae

Unfortunately, we were unable to identify the ancestor of the family Vangidae on the basis of our currently published results. The primary reason for this seems to be our failure to capture a member of the Prionopidae (helmet shrikes), a possible ancestor of the Vangidae, despite our efforts to visit Africa for that purpose. However, a previous phylogenetic study revealed that helmet shrikes are far more distant than the subfamily Malaconotinae (bush shrikes)[17]. Therefore, it is not probable that helmet shrikes would be included within the Vangidae. We are by no means entirely confident that the ancestor of the Vangidae is unrelated to helmet shrikes, though the present results suggest that the Vangidae has a closer relationship with the Cracticidae (bell magpies and Australian butcherbirds), rather than with the Laniidae. Thus, the assumption that the Vangidae is closely related with the Laniidae seems even more misplaced.

When did the ancestor of the vangas arrive on Madagascar? According to the theory of molecular clocks, the degree of mitochondrial DNA substitutions between taxa can be used to date evolutionary divergences. A rate of 2% nucleotide substitution per million years, referred to as the "standard clock", had been broadly applied based on enzyme fragment length data and the cytochrome *b* gene sequence in avian systematics. Following this assumption, it is inferred that the ancestor species of the vangas immigrated to Madagascar approximately 1.5 million years ago.

How did vangas diversify within Madagascar?

When we allocate the four species of *Newtonia* into the family Vangidae, we can recognize 19 species within the family. Since morphological differentiation within the family is extremely high, eleven (or twelve) genera are monotypic (Table 8-1), and the phylogenetic relationships within the family have been unsolved. Moreover, as depicted in Figure 8-5, we have failed to clarify the inter-specific relationships with high bootstrap values, which indicate the reliability of each cluster. We suspect that the principal cause of this failure might be the insufficient numbers of bases analyzed in our previous study. Therefore, we decided to continuously increase the range of 16S rRNA to approximately 1,500 base pairs and to compare the sequence once more in order to elucidate phylogenetic relationships within the family.

Figure 8-6 is a NJ dendrogram within the Vangidae using the Bull-headed

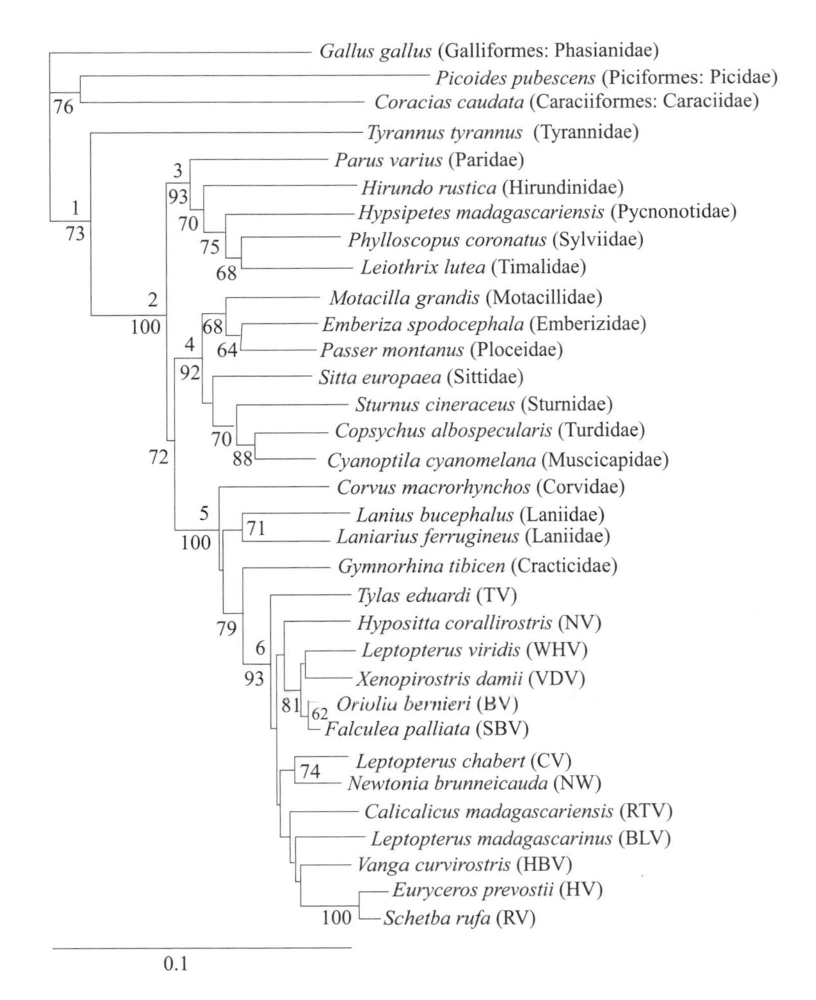

Fig. 8-5 A dendrogram derived from a 12SrRNA and 16SrRNA sequence (880 base pairs) by the neighbour-joining method. The vangids form a distinct cluster as Group 6. For abbreviated names and scientific names within the Family Vangidae, refer to Plate III-1 and Table 8-1, respectively. Numbers below branches indicates values of the bootstrap proportions in the 1000 repetitions (< 50 % are omitted). Bar equals 0.1 of Kimura's genetic distance (modified from Yamagishi et al. 2001[18]).

Shrike *Lanius bucephalus* and the Black-backed Magpie *Gymnorhina tibicen* as outgroups. This analysis could reveal the phylogenetic relationships within the family fairly clearly. Results of maximum-likelihood (ML) and maximum-parsimony (MP) analyses are consistent with the NJ dendrogram at the level of bootstrap proportions $\geq$70%. We thus recognized these consentient clusters as

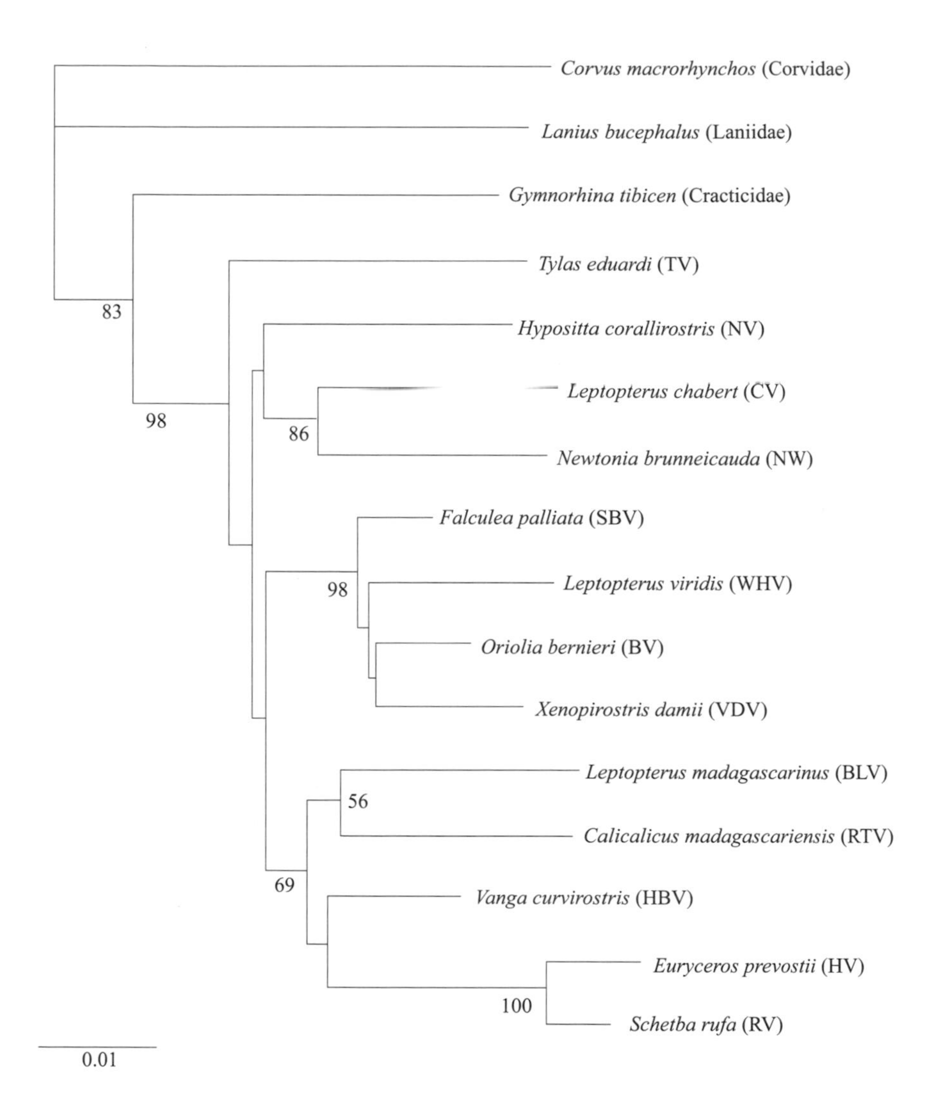

Fig. 8-6 A dendrogram derived from a 12SrRNA and 16SrRNA sequence (1437 base pairs) by the neighbour-joining method. For abbreviated names and scientific names within the Family Vangidae, refer to Plate III-1 and Table 8-1, respectively. Numbers below branches indicate bootstrap proportions in the 1000 repetitions (< 50 % was omitted). Bar equals 0.1 of Kimura's genetic distance (Yamagishi et al., unpublished).

valid, and depicted the phylogenetic relationships as shown in Figure 8-7. In the light of this figure, let us trace the evolutionary patterns of Rufous Vanga. It appears that within a relatively short period of time after having arrived on Madagascar, the common ancestor of the Vangidae simultaneously differentiated into five lineages. The five lineages are as follows: Group 1) Nuthatch Vanga; Group 2) Tylas Vanga *Tylas eduardi*; Group 3) the ancestor of Chabert's Vanga and the Common Newtonia; Group 4) the ancestor of Van Dam's Vanga, Bernier's Vanga *Oriolia bernieri*, Sickle-billed Vanga and White-headed Vanga; and Group 5) the ancestor of Red-tailed Vanga, Blue Vanga, Hook-billed Vanga, Rufous Vanga and Helmet Vanga. Since the divergences of these five lineages occurred simultaneously, their order cannot be ascertained from molecular data. In Madagascar, the absence of some other families seems to have left various niches vacant and available to the Vangidae. The rapid diversification of vangas is most likely to have occurred due to its utilization of the vacant niches normally occupied by other families.

As in the previous analysis, we estimated when this first divergence must have occurred, assuming that a rate of 2% nucleotide substitution per million years. On this occasion, however, based on 1,500 base pairs, we obtained a little different result from three million years ago. Usually, such a calibration was conducted using the cytochrome *b* gene in avian systematics, and there was no guarantee that it would agree with the data obtained using 12S and 16S rRNAs which include highly conservative regions in mitochondrial DNA. To ensure precise estimations, it is necessary to perform additional calibrations using the evidence of fossils, etc. Unfortunately, no fossils of the Vangidae have been recovered. Therefore, the correct answer, i.e., the estimated time, lies approximately between 3 and 1.5 million years ago.

The ancestor of Group 3 split into the Chabert's Vanga and the Common Newtonia approximately 2.3 million years ago, and that of Group 4 differentiated into Bernier's Vanga, Sickle-billed Vanga and the White-headed Vanga approximately 1.1 million years ago, after Van Dam's Vanga differentiated. Lastly, the ancestor of Group 5 initially differentiated into the Red-tailed Vanga, the Blue Vanga and the Hook-billed Vanga in that order, and then finally differentiated into Rufous Vanga and the Helmet Vanga approximately 800,000 years ago.

Among Groups 3, 4 and 5, a high level of similarity exists between the group consisting of Van Dam's Vanga/Bernier's Vanga/Sickle-billed Vanga/White-headed Vanga and that comprising Rufous Vanga/Helmet Vanga that differentiated relatively recently. Although there is great diversity in bill shape in each group (between Sickle-billed Vanga and White-headed Vanga, and between Helmeted Vanga and Rufous Vanga), there are close similarities in plumage coloration (Plate III-1). Vangas belonging to the former group exhibit black or white coloration while those of the latter group have liver and black coloration. The coloration resemblance is especially high between Sickle-billed Vanga and White-headed Vanga and between Rufous Vanga and Helmeted Vanga, though these all have different types of bills. In other words, the development of such variation in beak size and shape appears to have occurred in the relatively short time span of one million years. In contrast, variation in plumage coloration appears to have been more conservative and far more gradual.

By contrast, although Pollen's Vanga and Tylas Vanga are cxtremely similar in terms of plumage coloration (Plate III-1), they belong to different groups; Tylas Vanga belongs to Group 2, whereas Pollen's Vanga, whose blood we were unable to obtain, presumably belongs to Group 4, together with other congeners (Table 8-1). The precise reason for the close similarities in plumage coloration of two species belonging to different lineages remains unclear.

Additionally, we found another interesting feature. In the cases of Sickle-billed Vanga, White-headed Vanga and Van Dam's Vanga, it was observed that all these birds have an intense black coloration on the inside of their mouths, as if they had been painted with black ink.

Unfortunately, no information on this point is available in the case of Bernier's Vanga. However, if Bernier's Vanga also exhibits a black coloration on the inside of its mouth, this feature would be recognized as a synapomorphy of Group 4.

Eye color, on the other hand, does not reflect phylogeny. Among the vangas, four types of eye colors are known in the combination of periphery (iris) and center (pupil) (Plate III-1): both periphery and center are black (B/B), yellow and black (Y/B), red and black (R/B), and gray and black (G/B). However, as depicted in Figure 8-7, the eye color type is not related to phylogeny. It is noteworthy that the eye color of Red-tailed Vanga and Red-shouldered Vanga, which are subspecies and are considered to have a sister relationship, is different: the former is B/B type, while the latter is Y/B type (Plate III-1). Thus, eye color appears to be a characteristic that can be easily changed in the evolutionary process.

Thus the lineage within the Vangidae has been revealed using additional molecular data. We will now compare these findings with the taxonomy based on morphology. Most of the genera within the Vangidae are classified as monotypic, however three genera, *Leptopterus*, *Xenopirostris*, and *Newtonia*, contain a few species. Of these, taxonomists often changed the generic allocation of three species of the genus *Leptopterus* seusu lato, Chabert's Vanga, White-headed Vanga and Blue Vanga. The latter two species were sometimes assigned

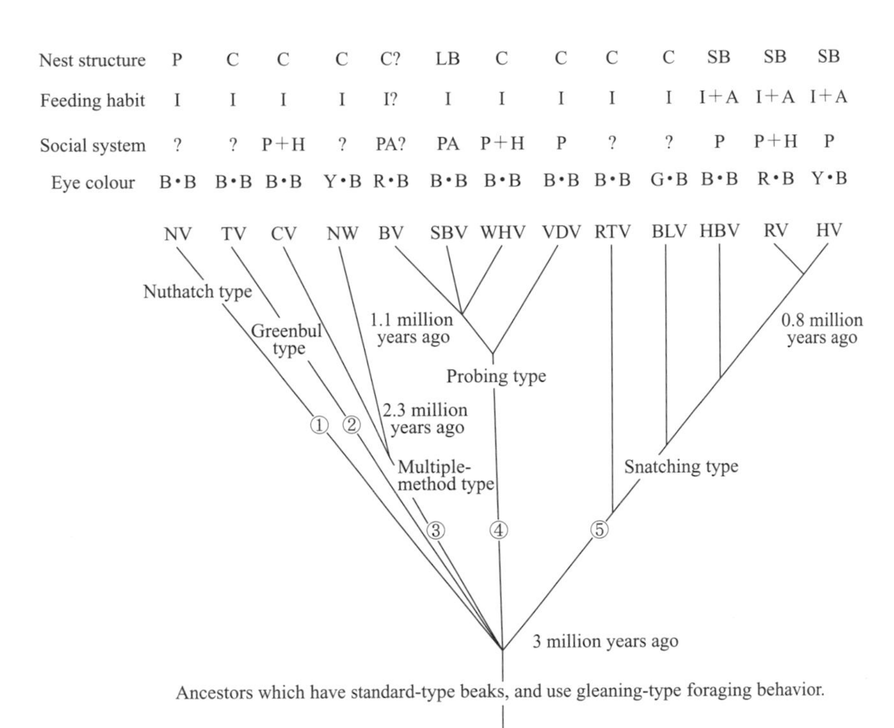

Fig. 8-7 Schematic diagram of characters and behaviors within the Family Vangidae. Nest structure; refer to Table 8-4 (Eguchi, unpublished). P: pile up, SB: small bowl, C: cup, LB: large bowl. Feeding habit; I: insect eater, I + A: insect + small animal eater. Social system; P: pair type, P + H: pair + helper type, PA: polygynandry type. Eye colour; B • B: black and black type, Y • B: yellow and black type, R • B: red and black type, G • B: grey and black type. For abbreviated names and scientific names, refer to Plate III-1 and Table 8-1, respectively.

into *Artamella* and *Cyanolanius*, respectively[19]. Our results clearly show that those three species belong to different groups (Fig. 8-7). This suggests that these three species differentiated at a relatively early period, and it also supports the taxonomic account that each of these three species should be assigned to different genera. In contrast, with regard to the relationships between Sickle-billed Vanga and White-headed Vanga, and between Rufous Vanga and Helmet Vanga, they exhibit extreme differences in bill shape. However, as mentioned above, the former and the latter seem to have diverged within Groups 4 and 5 respectively. Such relationships are consistent with the results by Schulenberg based on cytochrome *b* gene.

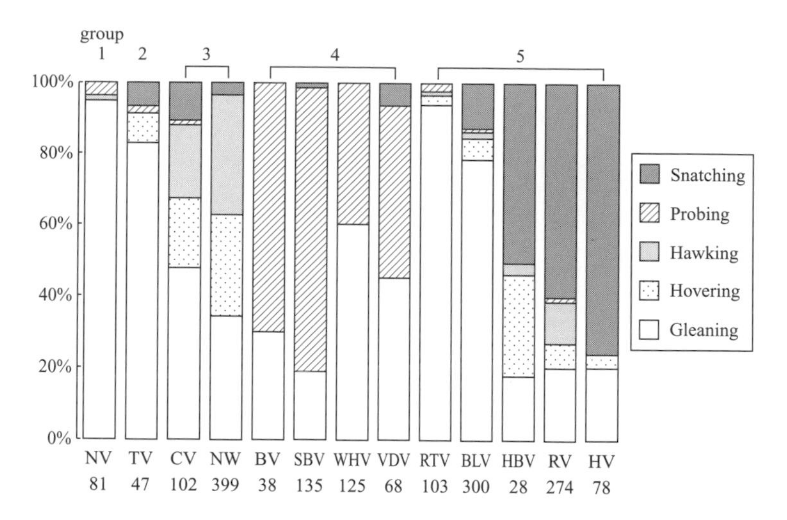

Fig. 8-8 Foraging behavior of vangids. Only the species of which lineages were proved in Figures 8-6 and 8-7 are listed here. The numbers below species names indicate number of observations. For abbreviated names and scientific names, refer to Plate III-1 and Table 8-1, respectively. On the basis of this figure, types of foraging behavior in Fig. 8-7 were determined (except NW and BV, from Yamagishi and Eguchi 1996[20]).

Phylogeny and foraging behavior

Further interesting implications can be derived (Fig. 8-8) by combining this phylogeny and the findings on foraging behavior obtained previously by Yamagishi and Eguchi[20]. All species of vangas capture prey by pouncing and picking, i.e., pouncing on sedentary prey items and picking them off–a behavior that in this book is termed as "gleaning". Among the vangas, although the extent to which they use these techniques varies, (1) all the species use a gleaning technique to capture prey. Moreover, (2) the older the lineage, the higher the ratio of birds that use the gleaning technique. Since vangas subsist mainly on insects, it is inferred that the ancestor of the vangas was an insect eater that used a gleaning technique, and had a bill of a standard type.

Members of Groups 1 and 2 primarily employ the gleaning technique. In Group 3, it is characteristic for birds to use various techniques in almost equal proportions. Group 4 is characterized by the use of a probing technique that involves poking and winkling out, as observed among woodpeckers. However, in Group 5, which differentiated at the earliest period, Red-tailed Vanga continue to primarily employ a gleaning technique. Yet in the case of Blue

Vanga, which differentiated next to Red-tailed Vanga, the suggestion of a snatching technique is evident. The later the time of differentiation, the greater the utilization of the snatching technique. As will be described in the next chapter, snatching is utilized to a great extent and the incidence of preying on small animals other than insects, such as lizards and chameleons, is high among Hook-billed Vangas, Rufous Vangas, and Helmet Vangas, which were the last species to differentiate[21,22]. Considering such foraging techniques, probing behavior is estimated to have evolved in Group 4, and snatching behavior seems to appear after the differentiation of Red-tailed Vanga but before that of Blue Vanga.

In summary, Rufous Vanga belongs to one of five groups that differentiated almost simultaneously immediately after the common ancestor of vangas arrived on Madagascar approximately three million years ago. Other than Rufous Vanga, this group includes Red-tailed Vanga, Blue Vanga, Hook-billed Vanga and Helmet Vanga. It is supposed that Rufous Vanga differentiated from Helmet Vanga approximately 800,000 years ago. This group is characterized by the use of a snatching technique in foraging, and it is considered that the group originally used a gleaning technique, as observed among Red-tailed Vanga. However, since the differentiation of Blue Vanga, this group began to pounce upon insects too, and since the differentiation of Hook-billed Vanga, this group began to pounce upon small animals, such as chameleons and lizards as well. Among them, Rufous Vanga is particularly characteristic in terms of pouncing upon prey on the ground[20].

Previous studies have sometimes tried to discuss ecological and behavioral evolution on the basis of molecular phylogeny. However, it is difficult to find evidence that behavioral features are closely related with their phylogeny. Perhaps that is because in species other than the vangas, birds had to adapt themselves by changing their behavior because they needed to establish themselves in an area where the ecological niche was already almost fully occupied. Conversely, it is inferred that the vangas did not have to change their behavior because the niche was virtually vacant, and the diversification of bill shape must have been a means of partitioning the foraging habitat and available prey. In comparison, their nest-building behavior and society is highly plastic.

Phylogeny, nest construction, and social structure

First, let us examine the nest construction of the vangas. In Chapter 4, it was mentioned that the nest of Rufous Vanga is bowl shaped, however this shape is most similar to the nest of Hook-billed Vanga. Consider the nest construction of vangas on the whole. Yamagishi had suggested that the nest construction of the vangas could reflect the phylogeny[23]. Subsequently, Eguchi produced Table 8-4 on 15 species of vangas on the basis of firsthand knowledge and literature sources. According to him, the vangas have four basic types of nest shape. (1) Sickle-billed Vanga build a large-sized, bowl-shaped nest (LB) like that of the Turtledove and the Crow, by piling up tree twigs. (2) Nuthatch Vanga build a nest by piling up moss in the hollow of a tree trunk (P). (3) Rufous Vanga, Hook-billed Vanga, and Helmet Vanga build small, bowl-shaped nests by piling up nesting material in the fork of a tree (SB). And (4) all the remaining species build cup-shaped nests (C), although the size and the location of the nest varies. Yamagishi combined (3) and (4) into "cup-shaped" nests and categorized the nests of the vangas into three types[23]. However, Eguchi regarded these nests as being rather bowl-shaped because the nests of the three species including the Rufous Vanga are too large to be termed "cup-shaped." Further, Eguchi considered that these nests are made by piling up nesting material and they also differ from the cup-shaped nests in that the nesting material appears to have been woven together.

Among the species that build cup-shaped nests, three species of the genus *Xenopirostris* and Chabert's Vanga fix their nests on horizontal branches, termed "fixing type" in this book, and they use advanced nest-building techniques, such as weaving plant fiber into a basket shape, fixing the nest on horizontal branches, etc. In comparison, other species that build cup-shaped nests, such as Tylas Vanga, Blue Vanga, Red-tailed Vanga and White-headed Vanga, build a hanging nest on the tip of a branch ("hanging type"), as does White-eye *Zosterops*. The nesting material used in the nests of Sickle-billed Vanga and Nuthatch Vanga differs greatly from that of other species. Moreover, Eguchi asserts that in the case of other species, excluding the two mentioned above, spider webs are a commonly used nesting material (Table 8-4).

It has been clearly established that the nest shape and nesting materials of swallows, martins and ovenbirds are related to their phylogenetic relationships[24,25]. Although the piled-up bowl diverged first, following the hanging cup and fixed cup, in that order, it appears that phylogeny and nest construction are in fact not closely related (Fig. 8-7). The three species, Rufous Vanga, Hook-billed Vanga and Helmet Vanga which are similar in nest shape, nesting material and location of nest, are also phylogenetically similar and form a monophyletic group (Table 8-4, Fig. 8-7). Red-tailed Vanga and Blue Vanga, which constitute Group 5 with the above mentioned three species, build a cup-shaped nest. However, the nest of Red-tailed Vanga is a fixed type and that of Chabert's Vanga is a hanging type. Sickle-billed Vanga is phylogenetically similar to White-headed Vanga and the genus *Xenopirostris*, but the nest shapes of these taxa are significantly different

Table 8-4 Morphological features, habitat, vegetation and structures of nests.

Japanese name	Genus	Body size (cm) /Habitat		Shape of nest	Location of nest	Nest material			
								Moss and lichen	Spider silk
Nuthatch Vanga NV	*Hypositta*	14	RF	piled up(P)	hollow of tree trunk			○	
Helmet Vanga HV	*Euryceros*	31	RF	small bowl(SB)	fork of tree	plant fibre	dead twig, (fernery)	○	(○)
Rufous Vanga RV	*Schetba*	20	RF/DF	small bowl(SB)	fork of tree	(plant fibre)	leaf stem	(○)	○
Tylas Vanga TV	*Tylas*	20	RF/DF	cup(C)	tip of branch	(plant fibre)	leaf stem	(○)	○
Bernier's Vanga BV	*Oriolia*	23	RF	cup?(C)	palm leaf *	plant fibre		(○)	
Hook-billed Vanga HBV	*Vanga*	29	ALL	small bowl(SB)	fork of tree	plant fibre)	leaf stem		○
Blue Vanga BLV	*Leptopterus*	16	RF/DF	cup(C)	tip of branch		leaf stem		○
Red-tailed-Vanga RTV	*Calicalicus*	14	RF/DF	cup#(C)	tip of branch		leaf stem		○
Van Dam's Vanga VDV	*Xenopirostris*	23	DF	cup(C)	on branch	plant fibre	grass blade		○
Pollen's Vanga PV	*Xenopirostris*	24	RF	cup(C)	on branch	plant fibre	grass blade		○
Lafresnaye's Vanga LV	*Xenopirostris*	24	SF	cup (C)	on branch	plant fibre	grass blade		○
Chabert's Vanga CV	*Leptopterus*	14	ALL	cup(C)	on branch	plant fibre			○
White-headed Vanga WHV	*Leptopterus*	20	ALL	cup(C)	on branch	(plant fibre)	dead twig		○
Sickle-billed Vanga SBV	*Falculea*	32	DF/SF	large bowl(LB)	fork of tree		dead branch, vine		
Common Newtonia NV	*Newtonia*	12	ALL	cup(C)	tip of branch	plant fibre	oliage		○

Habitat: RF = tropical rainforest, DF = deciduous broad-leaved forest, SF = semi-desert thorn forest, ALL = inhabit all three types of forest./= inhabit both types of forest.
Shape and material: cup #= roughly constructed, plant fibre = mainly rootlets, ( ) = the amount is small, *space in between bases of palm leaves.
References: Body size and habitat from Langrand (1990)[1], data of nests from Langrand (1990)[1], Rakotomanana et al. (2000)[21], Yamagishi(1994)[23], Thorstrom and de Roland (2001)[26], Putnam (1996)[27], Others are from Eguchi, unpublished.

(Table 8-4). Various nest types are observed among these species, such as cup-shaped nests (both fixed and hanging types) and nests that are piled up with tree twigs (Fig. 8-7).

Thus, among the vangas, nest shape, nesting material and nesting behavior reflect their phylogeny in certain cases but not in others. Eguchi presumes that this occurs because there is a portion of nesting ecology that reflects phylogeny, and there is another portion that can be changed easily because of ecological factors such as habitat and the availability of nesting material. For instance, Nuthatch Vanga and Helmet Vanga, which inhabit humid rainforests, build nests using plenty of moss, whereas Sickle-billed Vanga, White-headed Vanga and the members of *Xenopirostris*, which inhabit the relatively dry open forest, build nests using leaves and dead twigs (Table 8-4). If any alteration occurs in the divergence of vangas in the future, such evolution of nesting behavior will also have to be revised, and it would be necessary to analyze the nest construction of each species in a more qualitative manner.

As described in the next section, study on the comparative society of vangas has been primarily undertaken by M. Nakamura. His study has recognized three societies: pair society (P), cooperative breeding society accompanied by helper (P+H), and polygynandry (multiple husbands and multiple wives) society (PA). However, these societies do not reflect the five lineages in any manner. Each of these societies recurs within the same lineage (Fig. 8-7). We would prefer to let Nakamura study the determining factors of the social system.

Now that we have clarified the vangid phylogeny on the basis of molecular data, we can appreciate that the previous classification, established by taxonomists on the basis of morphology, has been largely accurate in some aspects. As to others, molecular phylogenetic analyses are useful tools to resolve taxonomic confusions. Especially in the Vangidae, molecular biological methods are the most useful tools to elucidate the evolutionary processes of morphological, ecological and behavioral features. Because this family exhibits such extreme diversities in these characteristics, most of its genera are recognized as monotypic. We believe that our new proposal, "*Newtonia* belongs to the family Vangidae," will stimulate scientists, such as taxonomists, morphologists, ecologists and ethologists to further investigate vangid phylogeny and biology in the future.

However, it is not yet known why the Vangidae did not adaptively radiate to include frugivory, nectivory and gramnivory. It is not an exaggeration to say that the true study on adaptive radiation of vangas begins henceforth. This is because the most significant aspect of ecological study is to determine the reason for the occurrence of adaptive radiation and to elucidate the mechanisms by which this was achieved.

References

1 Langrand, O. (1990) *Guide to the Birds of Madagascar*. Yale University Press, New Haven.

2 Peters, D. S. (1996) *Hypositta perdita* n. sp., a new avian species from Madagascar (Aves: Passeriformes: Vangidae). *Senckenbergiana Biologica* 76: 7-14.

3 Goodman, S. M., Hawkins, A. F. A. and Domergue, C. A. (1997) A new species of vanga (Vangidae, *Calicalicus*) from southwestern Madagascar. *Bulletin of the British Ornithologists'Club* 117: 5-10.

4 Sinclair, I and Langrand, O. (1998) *Birds of the Indian Ocean Islands*. Struik, Cape Town.

5 Rand, A. L. (1960) Laniidae. In: Myar, E. and Greenway, J. C. Jr (eds) *Check-list of birds*

of the world. vol IX. Cambridge: Mus. Comp. Zool. pp 309-364.

6 Rand, A. L. and Deignan, H. G. (1960) Pycnonotidae. In: Myar, E. and Greenway, J. C. Jr (eds) *Check-list of birds of the world.* vol IX. Cambridge: Mus. Comp. Zool. pp 221-300.

7 Snow, D. W. (1967) Paridae. In: Paynter, R. A. Jr (ed) *Check-list of birds of the world.* vol XII. Cambridge: Mus. Comp. Zool. pp 70-124.

8 Watson, G. R. Jr, Traylor, M. A. Jr. and Mayr, E. (1986) Sylviidae. In: Myar, E. and Cottrell, G. W. (eds) *Check-list of birds of the world.* vol XI. Cambridge: Mus. Comp. Zool. pp 3-294.

9 Delacour, J. (1932) Les Oiseaux de la misson zoologique Franco-Angro-Americaine a Madagascar. *L'Oiseau Revue Fr d'Ornithol.* 2: 1-96.

10 Rand, A. L. (1936) Distribution and habits of Madagascar birds. Summary of the field notes of the misson Franco-Angro-Americaine a Madagascar. *Bull. Amer. Mus. Nat. Hist.* 72: 143-449.

11 Bock, W. J. (1994) History and nomenclature of avian family-group names. Bull. *Amer. Mus. Nat. Hist.* No. 222, New York.

12 Swainson, W. and Richardson, J. (1831) Fauna boreali-americana. Part 2. *The Birds.* pp 523. London.

13 Dorst, D. (1960) Consideration sur les Passereaux de la famille des Vangides. In: Bergman, G., Donner, K. O. and Haarman, L. V. (eds) *Proceeding of XII International Ornithological Congress*, vol 1. Helsinki. pp 173-177.

14 Howard, R. and Moore, A. (1991) *A complete checklist of the birds of the world.* 2nd ed. Academic Press, London.

15 Appert, O (1994) Gibt es in Madagaskar einen Vertreter der Pirole (Oriolidae)? Zur systematischen Stellung der Gattung *Tylas. Ornithol. Beobach.* 91: 255-267.

16 Shimoda, C., Yamagishi, S., Tanimura, M. and Miyazaki-Kishimoto, M. (1996) Molecular phylogeny of Madagascar vangids: Sequence analysis of the PCR-amplified mitochondrial cytochrome *b* region. In: Yamagishi, S. (ed) *Social evolution of birds in Madagascar, with special respect to vangas.* pp 19-26. Osaka City University, Osaka.

17 Shulenberg, T. S. (1995) Evolutionary history of the vangas (Vangidae) of Madagascar. Chicago: Ph. D. Thesis. Univ. Chicago.

18 Yamagishi, S., Honda, M. and Eguchi, K. (2001) Extreme Endemic Radiation of the Malagasy Vangas (Aves: Passeriformes). *J. Mol. Evol.* 53: 39-46.

19 Sclater, W. L. (1924) *Systema Avium / Ethiopicarum. A systematic list of the Birds of the Ethiopian Region*, Part I, pp 304, Taylor & Francis, Londres.

20 Yamagishi, S. and Eguchi, K. (1996) Comparative foraging ecology of Madagascar vangids (Vangidae). *Ibis* 138: 283-290.

21 Rakotomanana, H., Nakamura, M., Yamagishi, S. and Chiba, A. (2000) Incubation Ecology of Helmet Vangas *Euryceros prevostii*, Which are Endemic to Madagascar. *J. Yamashina Inst. Ornithol.* 32: 68-72.

22 Rakotomanana, H., Nakamura, M. and Yamagishi, S. (2001) Breeding Ecology of the Endemic Hook-billed Vanga, *Vanga curvirostris*, in Madagascar. *J. Yamashina Inst. Ornithol.* 33: 25-35.

23 Yamagishi, S. (1994) Does the structure of birds' nests reflect their plylogeny? *Iden* 48: 4-5 (in Japanese).

24 Winkler, D. W. and Sheldon, F. H. (1993) Evolution of nest construction in swallows (Hirundinidae). *Proc. Nat. Acad. Sci. USA* 90: 5705-5707.

25 Zyskowski, K. and Prum, R. O. (1999) Phylogenetic analysis of the nest architecture of neotropical ovenbirds (Furnariidae). *Auk* 116: 891-911.

26 Thorstrom, R. and de Roland, Lily-Arison R. (2001) First nest descriptions, nesting biology and food habits for Bernier's Vanga, *Oriolia bernieri*, in Madagascar. *Ostrich* 72:

165-168.

27 Putnam, M. S. (1996) Aspects of the breeding biology of Pollen's Vanga (*Xenopirostris polleni*) in southeastern Madagascar. *Auk* 113: 233-236.

Comparative society of the family Vangidae

Masahiko NAKAMURA

Comparisons

In ecological studies, comparisons are often made among the societies of several species in order to discuss the evolution of society. We also discuss the issue of a certain society evolving after adapting itself to certain environmental conditions in terms of the relationship between these conditions and society. For instance, J. H. Crook made an interspecific comparison of the effects of environmental factors on the social structure, structure of the aggregative pattern and habitat, feeding habits, etc., of the Weaver Bird. Crook discovered that species inhabiting forests form pairs and protect the territory, spacing themselves out from others, whereas species inhabiting more open, savanna-like habitats form groups and undertake colonial breeding[1]. Unlike such a conventional method of interspecific comparison that treats species paratactically, the rapid development of molecular phylogenetics during the last 10 years or so allows present researchers to practice a new type of interspecific comparison. This method includes an additional component: organisms contain information on their phylogenetic history and relationships.

As described in the previous chapter, a remarkable morphological differentiation of body size and bill shape is present among the vangas of Madagascar; yet, biochemical evidence has indicated that they are a monophyletic group[2]. It has also been acknowledged that vangas differentiated into five groups almost simultaneously, immediately after their ancestor species, which is a sister group to the Black-backed Magpie, immigrated to Madagascar from continental South America three million years ago (Chapter 8). Therefore, vangas are ideal material for a study that considers phylogenetic relationships when making interspecific comparisons within their society.

Among the species of vangas, the Rufous Vanga has been studied the most quantitatively and for the longest period of time. A detailed explanation of Rufous Vanga biology is provided in Parts I and II of this book. The breeding unit of this species is a pair or a group comprising a single female and multiple males. In such a group, young males act as helpers and assist the breeding pair in territory defense, anti-predator mobbing, provisioning chicks, etc. As already mentioned in this book, such a breeding system is termed "cooperative breeding," and among approximately 9,000 avian species, only about 220 species (3% of all avian species) are acknowledged to be cooperative breeders[3]. Is this cooperative breeding system present among other vangas as well or is this breeding system specific to the Rufous Vanga within this group? The society of the Rufous Vanga will be studied in greater detail by comparing it with each society of vangas and ranking it in relation to the entire society. Few studies have addressed the comparative sociology of the vangas. With the exception of the Rufous Vanga, original data concerning the breeding life history and breeding system of each species of vanga is scarce and fragmentary. Collating the existing fragmental data available for each species, Dr Eguchi concluded that although vangas exhibit extensive differentiation in morphology and foraging behavior, their interspecific difference within the social structure is small[4]. However, I believed that if we conducted a thorough field research on the breeding system of each species, we might detect as significant a diversification within the vanga society as that observed in bill shape. To address this issue, I stayed at Madagascar for six months, from October to December in 1999 and 2000. During that period, I was able to obtain diverse information on eight species of vangas, almost half the total number of species, owing to the assistance of several people. However, the quantity of data varies with species. In this chapter, the breeding system is defined on the basis of data on role assignments of the pair male and female from the nest-building period to the nestling stage. The breeding system in eight species of vangas is compared using key words, such as "environment," "breeding system," and "phylogeny." Subsequently, the position of the cooperative breeding system of the Rufous Vanga within the overall social organization of vangas is examined.

Distribution of vangas

Vangas are distributed over almost the entire area of Madagascar, except for the central highlands that extend across the central region of the island. The highlands are at a high elevation, and the vegetation is steppe: shrubs are predominant, but most shrubs are distributed patchily and are small in size.

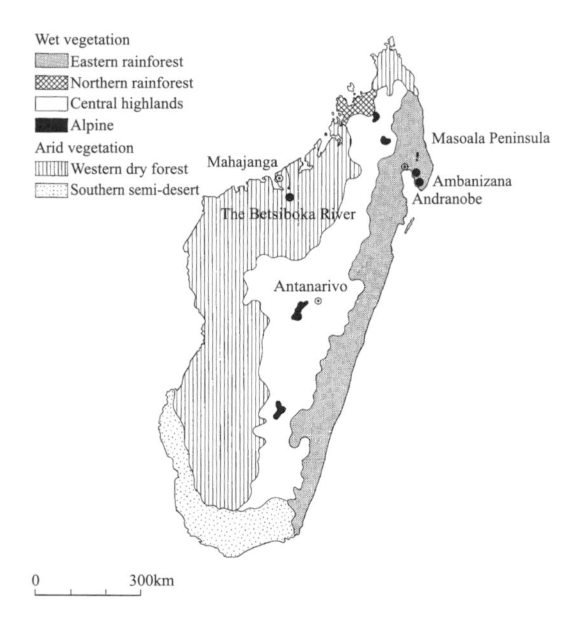

Fig. 9-1 Vegetation of Madagascar Island (modified from Sussman et al. 1985[5])

Thus, several arboreal birds, including vangas, are unable to survive in the central highlands. Instead, birds adapted to steppe habitats are predominant there.

Such an island-wide distribution does not imply that all types of vangas are distributed across all the corners of Madagascar Island except the central highlands. The climate of eastern Madagascar and that of western Madagascar is distinctly separated by the central highlands. The vegetation is also separated into a humid eastern region and a dry western region. Furthermore, the southern region has a semiarid climate (Fig 9-1). The humid eastern region is vegetated by tropical or mountainous rainforest, the dry western region by deciduous broadleaf forest, and the semiarid region by semiarid thorn forest that comprises plant species adapted to arid environments (Fig 9-1). The distribution of each species of vanga is closely related to and dependent on the climate and vegetation. For example, Bernier's Vanga and the Helmet Vanga are primarily distributed in the Masoala Peninsula, which lies in the northeastern part of Madagascar, whereas the Nuthatch Vanga and Pollen's Vanga are distributed in the rainforest that lies in the humid eastern region (Fig 9-2). In contrast, the distribution of the Sickle-billed Vanga is biased to the western part of the island. The Hook-billed Vanga and the White-headed Vanga can be found across the eastern and western parts of the island (Fig 9-2). Furthermore, the Tylas Vanga is chiefly distributed across the eastern part of the island, and its distribution on the western side is limited. Van Dam's Vanga and Lafresnaye's Vanga are

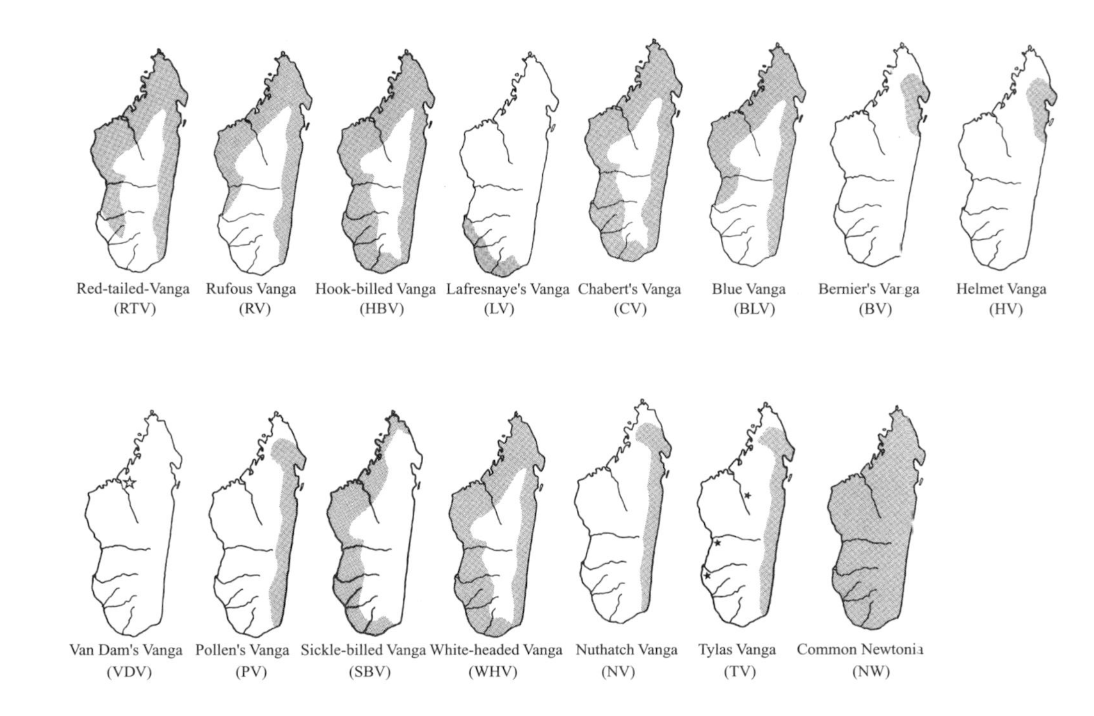

Fig. 9-2 Area of distribution of 15 species of vangids (cited from Langrand 19906). Capital letters within brackets represents abbreviations of names of species. Refer to Plate III-1 and Table 8-1)

distributed only locally in the dry western region. The Rufous Vanga, the Blue Vanga, and the Red-tailed-Vanga are extensively distributed over the eastern and western part of the island, excluding the area of semiarid thorn forest and the central highlands. Moreover, the Common Newtonia[2], which was newly assigned to the vangas by Yamagishi et al., is distributed throughout the island, unlike other vangas. However, in the central highlands this species is distributed locally (Fig 9-2).

Vangas are not migratory birds that seasonally migrate to the African and Eurasian continents. It has been reported that the Blue Vanga, which inhabits the broadleaf forest in the southwestern part of Madagascar, disappears from the forest during the dry season[7]. However, vangas are generally considered to be resident birds that inhabit the same habitat throughout the year and seldom migrate long distances[6].

Habitat of vangas

What then is the relationship between such a distribution and phylogeny? Dr Eguchi broadly classified the habitat of the vangas into four types: rainforest, broadleaf forest in the arid area, semiarid thorn forest, and all environments, which includes the previous three types[4]. Upon allocating these four habitats to each species on the molecular dendrogram mentioned in the previous chapter, it is found that vangas are not concentrated in a specific habitat. For instance, the Rufous Vanga inhabits both the rainforest and broadleaf forest, whereas the closely related Helmet Vanga inhabits only the tropical rainforest (Fig 9-3). Furthermore, among the four species of the Sickle-billed Vanga, Van Dam's Vanga, Bernier's Vanga, and the White-headed Vanga, the first two inhabit the dry broadleaf forest, whereas Bernier's Vanga inhabits the tropical rainforest, and the White-headed Vanga inhabits all environments. Consequently, not all the species that are phylogenetically closely related are concentrated in the same habitat (Fig 9-3). In other words, habitat use among vangas is not separated along phylogenetic lines, and several species coexist in each habitat.

In the transition zone from tropical rainforest to semiarid thorn forest, the average height of trees decreases, and simultaneously, crown density, density of the understorey, and density of creepers decrease. In other words, the number of layers within the forest decreases, and the foraging space used by each species diminishes. It has been acknowledged that this leads to the decreasing number of coexistent vanga species as the habitat undergoes a transition from tropical rainforest to semiarid thorn forest although multiple species coexist in each habitat[4]. Vangas primarily feed on insects, amphibians, and reptiles. They

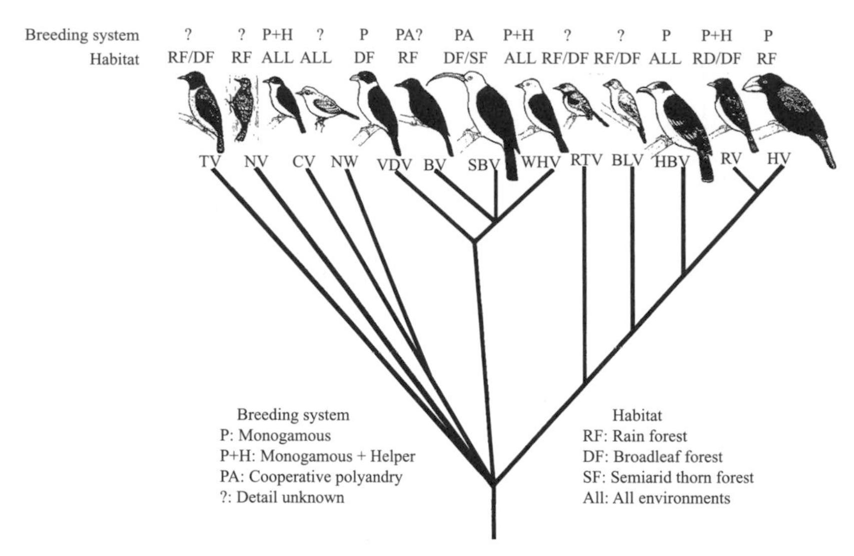

Fig. 9-3 Dendrogram showing relationships of 13 species of vangids, in relation to habitat, and breeding system (modified from Fig. 8-7. For abbreviated names, refer to Plate III-1 and Table 8-1)

coexist in each environment by employing species-specific foraging methods and by differentiating the foraging space that they adapt to.

Profiles of individual species

The breeding systems of birds can be broadly classified into five types: monogamy, polygyny, polyandry, polygynandry, and promiscuity. In monogamy, one male and one female form a pair, regardless of the length of the period, and in several cases, parents care for the eggs and chicks. In polygyny, one male copulates with several females. A male may copulate with several females at the same time (simultaneous polygyny) or sequentially (successive polygyny). In several polygynous breeding systems, the female performs most of the nestling tasks. Polyandry is the opposite of polygyny. In polyandry, one female copulates with several males at once (simultaneous polyandry) or in succession (successive polyandry). In several polyandrous breeding systems as well, most of the nestling care is undertaken by females. Polygynandry is a breeding system that combines polygyny and polyandry. In a group of multiple males and multiple females, males and females copulate with different individuals several times, and in most cases, both males and females care for the chicks. In promiscuous mating too, males and females copulate with different individuals several times. However, pair bonding is scarce, and in most species, the female alone cares for the eggs and chicks. These divisions are neither absolute nor rigid. The term "cooperative breeding," which refers to a system like that observed among the Rufous Vangas, applies to cases wherein individuals assist the related parents in chick rearing[9]. Moreover, cooperative breeding can be subdivided into several types on the basis of monogamy, polygyny, polyandry, etc. In addition, there is a type of monogamy wherein a pair acquires a territory and later disperses. In another type, multiple pairs undertake colonial breeding, and a breeding pair and the offspring in the colony breed cooperatively. Great variation exists in the cooperative breeding system of birds[10].

I would now like to introduce the breeding system of the eight vanga species for which I was able to collect data during this phase of the research. I will also describe a few events that occurred during the study.

The Helmet Vanga

The Helmet Vanga is smaller than a pigeon, and the male and female are of the same color-a black body, brown back and tail feathers, and a thick light-blue bill that appears disproportionately large in comparison with their body (Plate III-I). This bird inhabits the tropical rainforest in northeastern Madagascar and is relatively rare.

From October to November 1999, three research team members, Hajanirina Rakotomanana (nicknamed Haja), who is Malagasy, Mieko Kohno (then a graduate student of the Faculty of Science, Nagoya University), and I, conducted research on the Helmet Vanga at Andranobe and Ambanizana on the Masoala Peninsula, which lies in northeastern Madagascar (Fig 9-1). The Helmet Vanga was the first vanga species on which I had conducted comparative sociology studies; therefore, my expectations were high. However, shortly after we arrived at the airport of Maroantsetra, Kohno suffered from a bout of malaria, and we were stranded. From the outset, our research study in Madagascar was ridden with unpredictable events and challenges.

Shortly after we arrived at Andranobe, which is the first research field by boat from Maroantsetra, we visited the research station of the Peregrine Fund. Dr Russell Thorstrom, the American ornithologist, was working at this research station. Satoshi Yamagishi had visited Andranobe for research in October of the previous year, and he is an acquaintance of Dr Thorstrom. We explained to Dr Thorstrom that the objective of our research was to collect a small blood sample from the Helmet Vanga and conduct ecological research on this species and Bernier's Vanga. The blood sample collected from the Helmet Vanga was destined for molecular phylogenetic analysis. First, Dr Thorstrom told us that he had collected the blood sample of Bernier's Vanga, as requested by Yamagishi in 1998. He also informed us that he too was currently studying the Helmet Vanga and that he wanted us to refrain from conducting research on this species in Andranobe. However, he agreed to allow us to draw another blood sample and was kind enough to lead us to a nest that was in the incubating period. We set a mist net in the vicinity of the nest, caught the individual, and drew the blood sample-thus far, things had moved smoothly. However, an extremely unfortunate event occurred-the captured individual died in Haja's hand. All of us were completely shattered by this. Dr Thorstrom, who had shown us the location of the nest was particularly upset and was too disappointed to speak to anyone. I composed myself; nevertheless, it was with some difficulty that I was able to dissect the testes and stomach of the dead individual and preserve them in formalin.

The following day, we thanked Dr Thorstrom for his cooperation and apologized for having killed that precious individual. However, we felt like fleeing from Andranobe; therefore, we walked four hours to reach the second research field, Ambanizana, which was akin to skipping town. We were able to find one nest in Ambanizana, but there too, the research conditions were extremely harsh. From our lodging on the seashore, it took us two hours to reach the location of the nest. Two months prior to coming to Madagascar, I had injured the ligament of my right knee when I stumbled into a rock crevice while conducting research at Norikura Mountain in Japan. The two-hour climb to the study site, up the steep slope of the tropical rainforest proved to be extremely difficult for me since I had to drag my right foot, which was not healed completely. Finally, we arrived at the destination. However, we had no time to waste since we had to freshen up and immediately dress in raincoats and leggings in order to protect ourselves from the attacks of leeches and malaria-bearing mosquitoes. While we sat quietly on the ground for eight consecutive hours to observe the nest, leeches occasionally dropped off the trees, landed on our heads, and entered our ears. Insects swarmed around our body as well. We repeated our field research in such conditions for two days.

The total data we were able to obtain were based on two days of observations on the role assignment of the male and female during the incubation period, and on the testes removed from the individual we captured (or more accurately, killed) at Andranobe. Two individuals were participating in incubation. On observation through binoculars, we found that one of them was large in size and the other was smaller. From the measurement of specimens in the Botanical and Zoological Garden of Tsimbazaza, we already knew that the male of this species is larger than the female; therefore, we considered the larger of the two individuals involved in incubation to be male. We could barely identify between the incubating individuals on the basis of the slightly different colors on the tip of the tail feathers. We found that the clutch size was three (the clutch size of the nest at Andranobe was also three) and that the male and female, which were believed to be a pair, took turns at incubation. A single incubation period lasted 72.5 minutes for the male and 80. 0 minutes for the female, which is almost the same ratio. No other individuals were detected during the observation period[11]. On our return to Japan, we requested Dr Akira Chiba of Nippon Dental University to analyze the collected male testes. It was found that the testis was mature, but the wet weight was only 0.28% of body weight (103 g) and that the gonadal index was extremely low[11]. Testes weight is used to calculate the gonadal index, which is an important factor in inferring a bird species' breeding system. In general, the gonadal index is high in the case of polyandrous, polygynous, and monogamous colonial breeding species wherein there is a risk of extra-pair copulation. On the other hand, the gonadal index is low among monogamous species wherein a pair maintains a territory and is spatially separated from others[12]. In the case of the Helmet Vanga, the gonadal index is fairly low, and we considered this species to be monogamous since we observed both genders taking turns at incubating. At Andranobe, Dr Thorstrom told us that Helmet Vangas maintain territories and that incubation and nestling care are undertaken by both the male and female. Later, the research findings of Thorstrom et al. were published, and the content supported the idea that the Helmet Vanga is monogamous[13].

Incidentally, before I left Madagascar in 1999, Haja remorsefully informed me that he and I were notorious as bird killers (not murderers) at the Andranobe research station and that we would never be able to visit it again.

The Rufous Vanga (RV)

The Rufous Vanga is the main subject of this book. As the name suggests, the plumage of its back, wing, and tail are chestnut, and its stomach is white. The adult male is black from the head to the breast, and young males have black spots on a white background from the throat to the breast. The throat and breast of the female is whitish, and the chestnut color on its back appears dull. The body size of the Rufous Vanga is slightly smaller than that of the starling.

In this section, it is not necessary to explain the breeding lifestyle of the Rufous Vanga. Let it suffice to state that, as described in the other chapters, they follow the cooperative-breeding system founded around a monogamous pair. Most avian species are monogamous. Therefore, several cooperatively breeding species are established around monogamy, and the cooperative breeding observed among Rufous Vangas is a typical case.

The characteristic feature of Rufous Vangas that impressed me the most was the complexity of their vocal communication. They occasionally made alarm calls extremely similar to those of the closely related Helmet Vanga, and they also made a drumming sound by clicking their bills. I wondered whether this species had a "song."

A paper by J. Podos (2001) that recently appeared in the journal *Nature* explained that the adaptation of the bill of Darwin's Finches altered the song produced and that this led to the evolution of pre-mating isolation[14]. Other papers describing the song of Darwin's Finches were published in quick succession[14,15,16]. Society and vocal communication are believed to be closely interrelated. I wonder if the complexity of a species' society is directly related to the complexity of the vocal communication that it develops. As compared to the vocal sound of Darwin's Finches, that of the vangas is considerably more complicated; therefore, the analysis of interspecific comparison of the song of the vangas may be much more laborious. However, I believe that it would be worthwhile to consider this aspect of the vangas' social life when examining their social evolution and organization.

The White-headed Vanga

The gender of the White-headed Vanga can easily be ascertained in the field. The male is white from the head to the stomach and has a black back, wing, and tail feathers (Plate III-1). In comparison, the female is grayish from the head to the stomach. The White-headed Vanga is almost the same size as the starling, and as noted above, it is distributed across a wide range of habitats.

Since the White-headed Vanga was the next study target after the Helmet Vanga, I moved my study area from the tropical rainforest to the western arid forest at Ampijoroa and was surprised at how different these environments were. What struck me the most was the absence of any slopes at Ampijoroa! As compared to the two-hour climb upward to the research site in the tropical rainforest and the subsequent 1. 5-hour climb downward, the study site in the western arid forest was a plain, easily accessible, and conducive to research. Moreover, an accurate map of the study site had already been prepared. We also did not have to prepare our own meals. Most White-headed Vangas lived in the habitat nearest to the village rather than in the heart of the forest. Thus, elaborate outdoor gear was not required, and we wore simple sandals on our trip to the study area.

The territory of the White-headed Vanga is well-defined, and when we set mist nets near the nest and replayed the song of the male of the same species, the males of the territory flew into the mist net immediately in an effort to defend their territory. In comparison, females hesitated before flying in to defend the territory, but after several replays of the male's song, we were often able to capture females as well. As with the Helmet Vanga, we collected a small sample of blood from the vein beneath the wing of each captured bird, banded its leg, and then released it. In the case of the White-headed Vanga, the blood sample was used not only to analyze molecular phylogeny but also to assess parental relatedness and to assign gender through sex chromosome examination. Individuals that could not be identified by leg bands were identified by the subtle differences in plumage coloration. Thereafter, using a telescope or binoculars, we positioned ourselves at a distance of 10 m from the nest and observed the behavior of individuals for six consecutive hours from 05:00-11:00. The nest of the Helmet Vanga was located in the tropical rainforest, which has a comparatively cool environment; hence, we were able to continue our research there until the evening. At Ampijoroa, however, the afternoon sunlight was very harsh, and it was too hot to work. Therefore, we had to rest during the afternoons. Had we not rested and conserved our energy, we would have grown fatigued and become prone to sickness. Thus, we sat on the ground for six hours, observing the nest and noting data pertaining to aspects such as which sex of the bird visited the nest, the activities they performed, the work they did around the nest, etc. This observation method was employed by the late Kenzo Hada, my teacher at Shinshu University, in his study of the breeding life histories of various Japanese birds.

In contrast to the research on the Rufous Vanga, I did not have a specific study population in which members had been individually marked for several years. In a population wherein members were individually identifiable, basic data, such as birth history and breeding system, can be collected, analyzed, and interpreted. In my study, however, the only alternative, although a rather basic method, was to patiently observe each species' breeding pattern. That said, I do believe that this is a fundamentally important method in field biology.

In 1999, I found four nests, and I identified five individuals by leg banding and three individuals by plumage coloration. I examined the roles of the male and female from the nest-building to nestling stages and concluded that the White-headed Vanga was a typical monogamous species. My conclusion was based on the fact that during the nest-building period, the male and female contributed almost equally in the carrying of nest materials, and both the male and the female participated in incubation as well as in brooding and provisioning chicks (Fig 9-4). Moreover, the male had a distinct territory within which it mated with the pair female. Collectively, these observations seemed sufficient to define the breeding system of the White-headed Vanga as being monogamous. Upon my return to Japan after the field research in 1999, I immediately analyzed the data. On the basis of these findings, I prepared a paper and submitted it to a

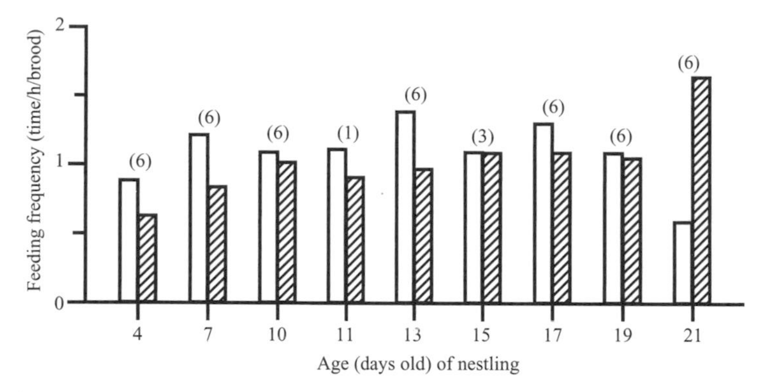

Fig. 9-4 Provisioning frequency of male and female White-headed Vangas (Open bars represent males; hatched bars represent females. Observation period (hours) each day are given in parentheses above the bars. Modified from Nakamura et al. 2001[17])

journal, which accepted it for publication. However, the breeding system of the White-headed Vanga has turned out to be less straightforward than it appeared.

During the 1999 research period, I was the only researcher studying this species. In such conditions, research efficiency is low. Therefore, it was necessary to include at least one collaborative researcher during the 2000 research period. Takayoshi Okamiya, a student who was pursuing a Masters' degree and had long been studying the Alpine Accentor with me at Norikura Mountain, was selected. I will never forget November 6, 2000, the day that Okamiya and I visited a nest during the incubation period.

My objective was to take a good photograph of the White-headed Vanga, and I set up the camera. As I was explaining to Okamiya that the White-headed Vanga was monogamous, I looked through the camera lens and noticed two individuals with female-like coloration engaged in some activity on the nesting tree along with the pair female, who had just left the nest after incubating. A closer look revealed that the pair female was preening with one of these individuals. It then preened with the other. Following this, all three birds flew to another tree. I thought my eyes were deceiving me, but I continued observing them. This time, the male, who had just finished incubating the eggs, hopped up the branches of the nesting tree and began alternately preening with the two female-like individuals. The pair female had returned to the nest after rotating incubation duties with the male; therefore, neither of these two individuals with female-like coloration was the pair female. What then were these two individuals?

Later research revealed that in two of the nine pairs examined, pair members were accompanied by individuals with female-like coloration: one pair had two such individuals, and the other had one helper individual. Along with the pair, these individuals participated in territory defense and in mobbing predators that approached the nest. However, they participated neither in incubation nor in brooding or provisioning the chicks[17]. Particularly in instances where potential predators, such as the Madagascar Coucal *Centropus toulou*, the Crested Coua, and the Brown Lemur, approached the nest, the male and the helpers mobbed the intruders more aggressively than the pair female who remained in the nest and relied upon the efforts of the breeding male and helpers. When we placed a mist net near the nest and replayed the song of a conspecific male, the pair male and the helpers reacted immediately. We captured one of the three helpers, leg banded it, and collected a blood sample. The helper was a young individual in the process of molting the plumage of the wing covers[17]. It is extremely important to ascertain the gender of the helper. In the case of most bird species, helpers are young males. Dr Isao Nishiumi of the National Science Museum utilized the CHD gene from the blood sample we obtained to ascertain that the helper was male.

On returning to Japan, I retracted the paper that I had written with such confidence a year earlier, wherein I stated that the breeding system of the White-headed Vanga is monogamous. I then rewrote the findings, stating that the breeding system of the White-headed Vanga is biased toward monogamy. However, a type of collaborative breeding occasionally occurs wherein helpers assist in territorial defense and anti-predator mobbing, but immature male helpers do not participate in provisioning the chicks. As already mentioned, in the territory of the Rufous Vanga, there exist males other than the breeding pair male, and some of these helpers assist the pair in its breeding efforts. A majority of these helpers are the yearling sons of the breeding pair and are sexually immature. Sons assist breeding pairs in territory defense, mobbing, and provisioning chicks. The degree of contribution of each son largely depends on the individual, and while some have a high contribution rate, others contribute little or not at all. The helpers of the White-headed Vanga were in molt. In the case of individuals undergoing molting, sex hormone secretion is not active, and such individuals are thus unable to breed[18]. On the basis of this evidence, the helpers of the White-headed Vanga were regarded as being unable to reproduce. Unfortunately, due to a shortage of samples, it was not possible to ascertain whether the immature male helpers observed in the case of the White-headed Vanga were sons of the breeding pair. However, the immature male individuals preened frequently with the breeding pair members in all the nests in which they were present. Thus, it would not be particularly surprising to discover that they are, indeed, related.

Van Dam's Vanga

Van Dam's Vanga is a stocky bird, slightly larger than a starling and with a thick black bill (Plate I-10). The male is black from head to the face, whereas the female's head is black only above the eyeline. Therefore, it is easy to distinguish between the sexes in the field (Plate III-1). This species is the rarest of all vangas and is extremely limited in distribution. It is known to occur in only one location- the northwest part of the island that is Ampijoroa!

Studies on Van Dam's Vanga were undertaken in Ampijoroa by Taku Mizuta at the same area where the Rufous Vanga was studied. We expected to easily capture Van Dam's Vanga by replaying the male's song, as we had successfully done with other vanga species. However, although Van Dam's Vanga did approach the speakers to inspect the source of the sound, it did not fly into the net, and ultimately, we were unable to leg band any specimens. However, we were able to identify individuals by subtle differences in the coloration around the face area. Thus, using this means of identification, we conducted our research on the task allocation of males and females.

At the Jardin A research area at Ampijoroa, we recorded the presence of six pairs of Van Dam's Vanga, including one pair that was not observed directly but was recognized by its song. At this location, the population density of Van Dam's Vanga was very low as compared with the population density of the Rufous Vanga. Therefore, the territory of each pair did not overlap that of other pairs and no interference between neighboring pairs was observed. We located three pairs in which nesting was underway and observed their nest-building, incubating, and chick-rearing behavior. The male and female contributed equally to all these activities (Fig 9-5), and no helper that assisted in breeding was

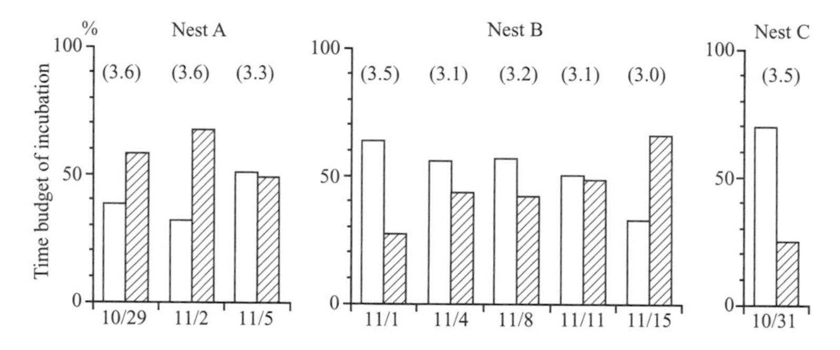

Fig. 9-5 Incubation time budgets of male and female Van Dam's Vanga (Vertical axis shows the rate of incubation time that male and female shared within total observation time. Open bars represents males; hatched bars represents females. Observation period (hours) is given in parentheses. Modified from Mizuta et al. 2001[19])

present[19]. Pairs had dispersed home ranges, and the two pair individuals cared for the eggs and chicks. Accordingly, we believe that Van Dam's Vanga has a monogamous breeding system.

The Sickle-billed Vanga

This species is the largest among the vangas, and it has a distinctive large, curved, gray bill, similar to that of a snipe. This bill shape is diametrically opposite to the thick and large bill of the Helmet Vanga (Plate III-1). The Sickle-billed Vanga noisily chirps "ga-ga," but occasionally chirps "kwa-kwa," fairly similar to the caw of a crow. In Malagasy, this bird is called voronjaza, a name meaning "bird of the children." It is so named due to the similarity between its "kwa-kwa" call and the crying of an infant.

During the first research on this species in 1999, I was able to locate four nests of the White-headed Vanga but only two nests of the Sickle-billed Vanga. Fortunately, however, two of seven individuals involved in breeding activities within the study area had previously been leg banded by Dr Eguchi in 1996, and I was also able to identify other individuals by their unique plumage patterns. Therefore, I was able to conduct accurate research. During my initial observations of this bird, there were several surprises. First, I was surprised to observe that in two of our study nests, multiple males and only a single female were present. I was able to conduct research from the nest-building stage to the fledging stage on the first nest that I found. However, the remaining nest was abandoned due to some unknown reason just before hatching. In the nest that I was able to observe until fledging, one female and three males were involved in breeding, and in the nest that failed, two males were involved with one female. The individuals involved in breeding were not helpers, as indicated by the fact that the female in these cases mated with all the males.

The courtship of the Sickle-billed Vanga is unique. Among birds, it is a common practice for the male to exhibit courtship behavior toward the female. However, in the case of the Sickle-billed Vanga, this relationship is reversed, and it is the female that courts the male (s). Upon finding a male, the female approaches it and then holds its own body in a horizontal posture, quivering both wings and tail feathers in a courtship display. During this time, the male does not exhibit any courtship behavior toward the female. Moreover, the mating period is also unique. Generally, courtship behavior among birds is restricted to the period from nest building until egg laying, i. e., only during the period when the female is receptive. In the case of the Sickle-billed Vanga, however, mating takes place not only during this period but also continues through the incubation period and into the middle of the chick-rearing period[20]. Additionally, copulation also continues until the middle of the chick-rearing period.

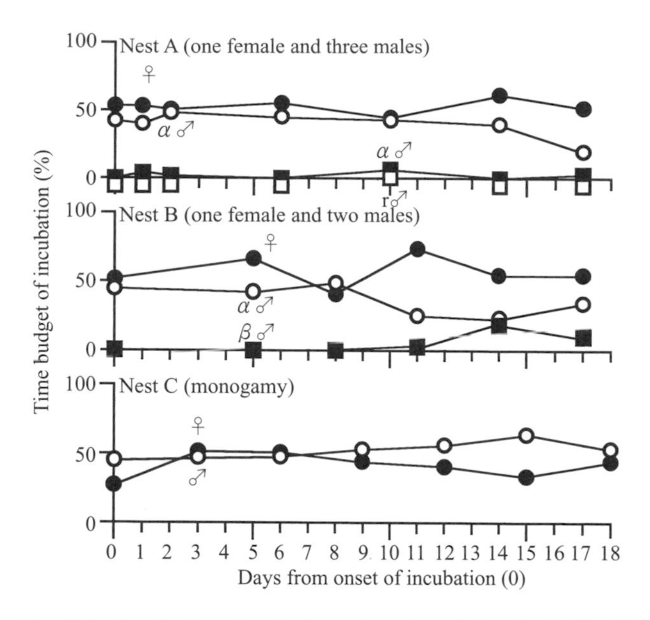

Fig. 9-6 Incubation time budget of male and female Sickle-billed Vanga
(Modified from Nakamura et al. 2001[20])

On the basis of my observations during the 1999 research period, I had already judged the type of breeding system present among the Sickle-billed Vanga; therefore, in 2000, I concentrated my research efforts on the Sickle-billed Vanga in order to collect as much information as possible along with Okamiya. We found that among the Sickle-billed Vanga, there are monogamous as well as polyandrous breeding units: out of seven nests that we located, three were monogamous and four were polyandrous (two males in three nests, three males in one nest)[20]. Moreover, a clear hierarchy existed among the males, with the highest-ranking alpha-male monopolizing copulations with the accompanying female during the week before the commencement of the incubation period. However, after the incubation period, females exhibited courtship only toward lower ranked beta- and gamma-males. Although there was a difference in the frequency, each male participated in incubation activities along with the breeding female (Fig 9-6), and low-ranked males contributed toward chick rearing more actively than the alpha-male (Fig 9-7).

The breeding system of the Sickle-billed Vanga is clearly different from that of the Rufous Vanga. In the former case, all helpers are breeding males, a situation that occurs because the female copulates with all male helpers. The breeding system in which more than two males copulate with the female and all

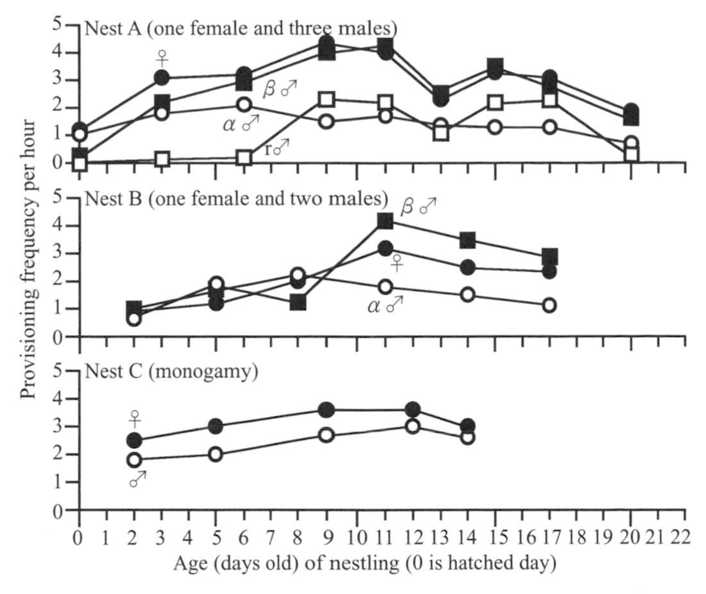

Fig. 9-7 Provisioning frequency of male and female Sickle-billed Vanga (Modified from Nakamura et al. 2001[20])

males assist in rearing the chicks is termed "cooperative polyandry." To date, cooperative polyandry has been known to exist in only nine bird species, and in all these cases, the male's sole contribution is toward provisioning the chicks[22]. However, among the Sickle-billed Vangas, the males assist in nest building, incubation, and chick rearing. In these respects, the Sickle-billed Vanga is clearly different from the other species that have thus far been known to practice cooperative polyandry.

The Hook-billed Vanga

The Hook-billed Vanga is larger than a starling and has a strong, hook-shaped bill, similar to that of a shrike. The male and female have identical coloration. A black band runs from the eye to the back of the head. The cap of the head and the body below the band is white (Plate III-1). The sound of their song is "hi-hi," similar to that of White's Ground Thrush.

The study on the Hook-billed Vanga was undertaken by Haja. We found one nest of this species in the tropical rainforest at Ambanizana and four nests in the arid forest at Ampijoroa. The male and female Hook-billed Vanga have identical plumage coloration. Thus, it is difficult to distinguish between them in the field. However, there did exist some individuals that did not have a hooked bill.

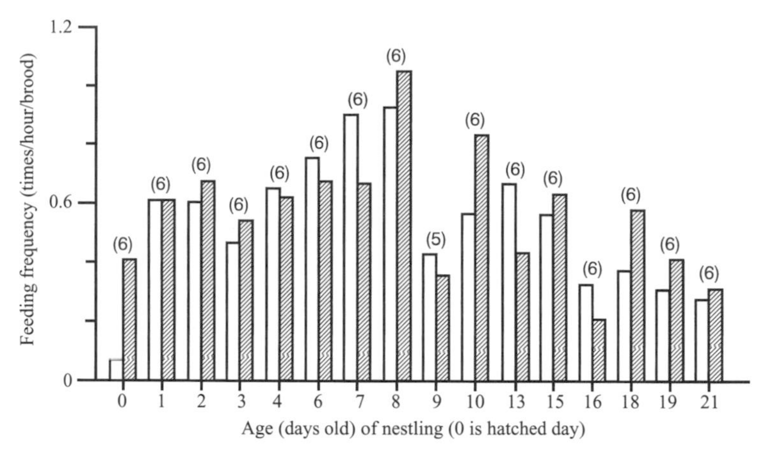

Fig. 9-8 Provisioning frequency of male and female Hook-billed Vangas (Open bars represents males; hatched bars represents females. Observation period (hours) is given in parentheses. Modified from Rakotomananaet et al. 2001[21])

Besides, the normally black first row of flight feathers of certain other individuals was turning white. We were unable to leg band any birds, but we were able to identify individuals on the basis of these individual differences in plumage coloration and body shape. However, we were unable to ascertain gender.

At Ambanizana, we also conducted research on the Helmet Vanga. However, due to time limitations, our research on the nest of the Hook-billed Vanga lasted for merely three days during the incubation period. At Ampijoroa, we were able to conduct continuous research on one nest from the nest-building to the chick-rearing stages, on another nest from the nest-building to the incubation periods, on a third only during the incubation period, and on a fourth during the chick-rearing period. Although the information obtained was necessarily fragmentary, we were able to document that individuals inferred to be members of a pair contributed equally to nest building, incubation, and chick rearing (Fig 9-8) and that helpers were not present at any of the studied nests.

We were able to determine the clutch size in only one nest at Ambanizana; it was located at a height of 10 m in a tree, as were the other nests of this species. Unable to climb to this height ourselves, we solicited the assistance of a young local villager, skilled at tree climbing, to check the nest contents. As a reward, we provided him with rice, a staple food in Madagascar. Haja assured me that when provided with the incentive of a large amount of rice, no feat was impossible for a Malagasy. The young man gave up trying to climb the tree on which the nest lay. Instead, using a creeper as a rope, he climbed up a tree that was five meters away and swung, Tarzan-like, back and forth, close to the nesting tree, peering into the nest each time he swung past. This unbelievable feat was performed for our benefit. The young man was unable to check the nest contents in one swing. Hence, he repeated the spectacle five times and checked the number and color of the eggs. The nest contained three eggs.

In general, among monogamous species, there is neither a significant difference between the body size of males and females nor any difference in plumage coloration and body morphology[22]. We compared the dimensions of the bodies of males and females of the Hook-billed Vanga using specimens and dead bodies that we collected. However, we were unable to detect any difference between males and females. Further, we suspected this species to be monogamous on the basis of our observations that two individuals participate in the breeding activities.

Bernier's Vanga

To be frank, I have never seen a Bernier's Vanga. This species is very rare and inhabits only a limited area of the tropical rainforest. To date, the male has not been photographed and data on the breeding ecology of this species is very scarce. Bernier's Vanga is smaller than the Helmet Vanga and of almost the same size as Van Dam's Vanga. The body of the male is entirely black like that of a crow (Plate III-1), but the body of the female is brown with black stripes on the belly.

At the research station at Andranobe, Dr Thorstrom, who kindly assisted us in our research on the Hemet Vanga, donated a blood sample taken from Bernier's Vanga. He also showed us a half-completed draft of a paper he had written on this species. Flicking quickly through the pages of his draft, he narrated how he had climbed 14 m up a tree to collect this sample from a nest. He explained that on the basis of his observation that incubation was undertaken by both the male and female, he believed Bernier's Vanga to be monogamous. Yet, in the paper that he subsequently published, he reported that in one nest out of the four studied, a young male was present in addition to the pair male, and it copulated with the pair female and assisted in incubation[23]. Its copulation with the female reveals the young male to be an extra-pair male rather than a helper. Thus, it is highly probable that the breeding system of Bernier's Vanga is that of cooperative polyandry, as observed in the case of the Sickle-billed Vanga.

Chabert's Vanga

This species, which is smaller than a sparrow, is among the smallest of the vangas. Like the Madagascar Paradise Flycatcher, the skin surrounding the eyes is exposed and is pale blue in color. Although it resembles a member of the flycatchers of the Old World, its beak is thicker (Plate III-1).

Appert (1970) described the basic breeding biology of the vangas, such as nest shape, clutch size, etc. He reported that in the case of Chabert's Vanga, individuals other than the pair members participated in provisioning the male that was incubating[24]. Prior to our study on the Rufous Vanga, this had long been the sole piece of information suggesting the existence of helping behavior among vangas.

In reality, the breeding system of Chabert's Vanga remains unknown. They forage actively on the tips of branches, high in the forest canopy. Their nest is located high in a tree, and they seldom descend any lower. Accordingly, it is difficult to capture and individually identify them. Moreover, I believe that even if we leg banded them, it would be very difficult to identify individuals in the field since the male and female have identical coloration and cannot be easily differentiated due to their rapid movements. Although Appert stated that the helper is female, I was unable to distinguish its gender.

However, we do have some fragmentary data on this species. During the chick-rearing period, I observed three adult birds alternately carrying food to the chicks at a nest. Further, in another nest, I observed three adult birds feeding one fledging chick. Similarly, during the 1999 research period, Satoshi Yamagishi and Dr Masami Hasegawa of Toho University observed multiple individuals carrying food to the nest during the chick-rearing period. Since numerous researchers independently observed that several individuals were involved in breeding, it is certain that Chabert's Vanga is a cooperative breeder. However, fundamental information, such as the nature of the helper's involvement during the period from nest building to chick rearing and the gender of the helper, remains unknown. Moreover, it is also possible that the helper might be a male that copulates with the female, as observed among the Sickle-billed Vangas. Ultimately, we will be unable to elucidate in detail the cooperative breeding pattern of Chabert's Vanga unless we individually identify birds by leg banding or marking, perform gender testing using blood samples, and undertake rigorous field research.

The relationship between phylogeny and the diversity of the breeding system

Differentiation within the vangas is most apparent in their morphology and foraging behavior. However, in terms of their social structure, differences at the interspecific level are thought to be marginal. Vangas speciated sympatrically in the forest habitat and did not expand to colonize other habitats beyond the forest[24]. Dr Eguchi has speculated that although the habitat of the vangas is structurally rather diverse, the width of the habitat niche is limited and the width of the feeding niche is also relatively narrow, and therefore, this did not result in large differences in the social structure of group members[4]. When he posited this hypothesis in 1995, the only study of the social ecology of vangas was that done on the Rufous Vanga. Thus, it is but natural that he proposed this hypothesis.

Since all of the breeding systems of all vanga species have not been elucidated, it is premature to arrive at any definitive conclusions. Nevertheless, it is clear that the breeding system of vangas is not uniform. Vangas are basically monogamous, but they exhibit several variations in cooperative breeding, and there exists considerable diversity in their breeding system. The White-headed Vanga is a cooperatively breeding species wherein immature molting males contribute to the breeding efforts of the pair by participating in territory defense and mobbing behavior against predators. As noted earlier, the hormonal secretion is inactive in molting individuals. Hence, they are unable to breed. While the White-headed Vanga's helpers are unable to participate in breeding activities, such as provisioning, because they are molting, they are involved in a close sociological relationship with pair members through preening activities. Among the Rufous Vangas, helpers do not undergo molting of the wing feathers and they have begun participating in the pairs' breeding activities by provisioning chicks. Thus, the cooperative breeding pattern of the White-headed Vanga lags a step behind that of the Rufous Vanga. At Ampijoroa, Asai and Yamagishi observed an individual Blue Vanga, which had not yet completed molting, in the company of a breeding pair. Could it be that this species also has a breeding system similar to that of the White-headed Vanga? The Sickle-billed Vanga and Bernier's Vanga form a breeding group consisting of a single female and multiple males. The Rufous Vanga also falls in the same category. However, in the case of the Sickle-billed Vanga and Bernier's Vanga, multiple males copulate with a single female. These are breeding males, not helpers, and the breeding system is that of cooperative polyandry. It is certain that Chabert's Vanga is also a cooperative breeder.

Is cooperative breeding concentrated within a specific taxonomic group of vangas? For example, although the Helmet Vanga is monogamous, the closest relative, the Rufous Vanga, is a cooperative breeder (Fig 9-3). On the other hand, the phylogeny and interrelationships are relatively well understood among Bernier's Vanga, Van Dam's Vanga, and the White-headed Vanga, and except for Van Dam's Vanga, these species are cooperative breeders. The degree of phylogenetic constraints on the breeding system remains unclear.

The most famous case of adaptive radiation among birds is that of Darwin's Finches of the Galapagos Islands. Using mitochondrial DNA from 13 species of Darwin's Finches and from several finch-like species of small birds distributed on the South American mainland, Sato et al. analyzed the genetic relationships among these species. They found that, as Darwin inferred, the group of Darwin's Finches is monophyletic and originates from a single ancestral species (closely related to the group *Tiaris*, which is widely distributed in South and Central America)[25]. The ancestral species is believed to have immigrated to the Galapagos Islands from continental South America over two million years ago. (Vangas represent an even earlier radiation.)

In the case of Darwin's Finches, as in the vangas, the bill shape differs among species, and this difference is reflected in varied feeding habits. It is known that within the population, the average size of the bill changes because of the drought-causing effects of El Nino, which occurs at intervals of several years. After these droughts, the female mortality rate increases and the population becomes biased toward males. The breeding style of Darwin's Finches is primarily monogamous; however, when an extreme sex-ratio bias occurs, it shifts to polygyny. However, unless such a sudden extreme climatic change occurs, Darwin's Finches are monogamous (although the breeding systems of all these species have not been clarified). Incidentally, even during the limited period of this research project, the Rufous Vanga, the White-headed Vanga, the Sickle-billed Vanga, Bernier's Vanga, and Chabert's Vanga were discovered to be cooperative breeders. Therefore, vangas are far more diversified in their breeding system than Darwin's Finches. However, the reason for the evolution of such a diversified breeding system is as yet unknown.

Breeding system and the environment

What then is the relationship between the environment and the breeding system? Here again, I refer to the data on breeding system presented in Figure 9-3. The Rufous Vanga is a cooperative breeder that inhabits the deciduous broadleaf forest at Ampijoroa, whereas its closest relative, the Helmet Vanga, inhabits only tropical forests and has a monogamous breeding system (Fig 9-3). Similarly, the Hook-billed Vanga, which inhabits the tropical forest, is monogamous. Although not included in this dendrogram, Pollen's Vanga, which inhabits tropical rainforests, also appears to be monogamous[27]. It is highly probable that Bernier's Vanga, which inhabits the tropical rainforest, is cooperatively polyandrous. However, the Sickle-billed Vanga, which inhabits the dry thorn forest and the deciduous broadleaf forest, is a cooperatively polyandrous breeder, and the White-headed Vanga, which inhabits all environments and breeds in the deciduous broadleaf forest, is a cooperative breeder (Fig 9-3). Chabert's Vanga is also a cooperative breeder. It appears that the breeding system is most strongly influenced by the breeding environment of each species rather than by phylogeny. If we pursue this theme to its logical conclusion, we would predict that the Rufous Vanga populations inhabiting the tropical rainforest should not be cooperative breeders but should have adopted a monogamous breeding system, and that tropical forest-dwelling populations of Chabert's Vanga and the White-headed Vanga (which occur in all habitats) should have adopted a monogamous breeding system.

To what extent does the environment influence the regulation of the breeding system? As the initial step in the study of the comparative society of vangas, I decided to first examine the general breeding system of each species and then attempt to correlate each species' breeding system with the habitat occupied. However, this approach ignores the mechanisms whereby the life history style and habitat of each species regulates the breeding system. This issue should be borne in mind when planning future research endeavors on this subject.

Conclusion

Unfortunately, my participation in this project lasted merely two years, and I was able to spend a total of only six months in the field. During that period, I attempted to observe as many species of vangas as possible and to collect as much diverse information as was feasible. In this chapter, I have covered only eight of the 15 species of vangas. It is only after the breeding systems of all the vanga species, including the Common Newtonia, have been examined using molecular tools and after the social system during the non-breeding period has also been elucidated, that the hypotheses presented in this chapter can be confirmed or refuted. Crucially, the fact that social organization and interactions take place not only during the breeding period but also during the non-breeding period should be acknowledged. Although I have not mentioned it in this chapter, it is noteworthy that during the non-breeding season, the Sickle-billed Vanga forms large flocks of 20-30 birds or more, and it forms large communal roosts of over 50 individuals at night[28]. Chabert's Vanga and the White-headed Vangas are also present in this communal roost, but the Rufous Vanga is never present. Among several birds, the society formed during the non-breeding period determines the society observed during the breeding period. Henceforth, we must focus our efforts on determining the breeding system of all vanga species. Thereafter, it would be necessary to review the society during the non-breeding period. Only then can the similarities between the society of the Rufous Vanga and the society of all other vangas be revealed. Thus, much remains to be accomplished.

Each time I observed vangas in Madagascar, I wondered what observations and insights Darwin would have derived with regard to the evolution of vangas, had his voyage on the Beagle led him to Madagascar instead of the Galapagos Islands. Although the Beagle sailed from Mauritius to Cape Town, the bow of the ship was unfortunately not directed toward Madagascar. What ideas or insights would he have had on morphological change, sympatric speciation, adaptive radiation, and the diversification of the social system, and what conclusions would he have drawn? He might have arrived at a completely different theory of evolution. It appears that my research in Madagascar, the laboratory of evolution, appears set to continue for several more years or even decades.

References

1 Crook, J. H. (1964) The evolution of social organization and visual communication in the weaver birds (Ploceinae). *Behaviour Supplement* 10: 1-178.

2 Yamagishi S., Honda, M., Eguchi, K. and Thorstrom, R. (2001) Extreme endemic radiation of the Malagasy vangas (Aves: Passeriformes). *J. Mol. Evol.* 53: 39-46.

3 Stacey, P. B. and Koening, W. D. (1990) Introduction. In *Cooperative Breeding in Birds* (eds. Stacey, P. B. and Koening, W. D.), pp. 9-18. Cambridge University Press, Cambridge.

4 Eguchi, K. (1995) Community structure and adaptive radiation of Madagascar birds. *Jpn. J. Ecology* 45: 259-275. (in Japanese).

5 Sussman, R. W., Richard, A. F. and Ravelojaona, G. (1985) Madagascar: Current Projects and Problems in Conservation. *Primate Conservation*, pp. 53-59.

6 Langrand, O. (1990) Guide to the Birds of Madagascar. Yale University Press, New Haven.

7 Appert, O. (1968) La repartition geographique des Vangides dans la region du Mangoky et al question de leur presence aux differntes epoques de l'annee. *L'Oiseau et la Revue Française D'Ornithologie* 38: 6-19.

8 Yamagishi, S. and Eguchi, K. (1996) Comparative foraging ecology of Madagascar vangids (Vangidae). *Ibis* 138: 283-290.

9 Emlen, S. T. (1991) Evolution of cooperative breeding it birds and mammals. In *Behavioural Ecology* (eds. Krebs, J. R. and Davies, N. B.), pp. 301-337. Blackwell Scientific Publications, Oxford.

10 Brown, J. L. (1987) *Helping and Communal Breeding in Birds: Ecology and Evolution.* Princeton University Press, Princeton.

11 Rakotomanana, H., Nakamura, M., Yamagishi, S. and Chiba, A. (2000) Incubation Ecology of Helmet Vangas *Euryceros prevostii*, which are endemic to Madagascar. *J. Yamashina Inst. Ornithol.* 32: 68-72.

12 Møller, A. P. (1991) Sperm competition, sperm depletion, parental care and relative testis size in birds. *Am. Nat.* 137: 882-906.

13 Marca, G. L. and Thorstrom, R. (2000) Breeding biology, diet and vocalization of the Helmet Vanga, *Euryceros prevostii*, on the Masoala Peninsula, Madagascar. Ostrich 71: 400-403.

14 Podos, J. (2001) Correlated evolution of morphology and vocal signal structure in Darwin's finches. *Nature* 409: 185-188.

15 Grant, P. R., Grant, B. R. and Petren, K. (2000) Vocalizations of Darwin's finch relatives. *Ibis* 142: 680-682.
16 Ryan, M. J (2001) Food, song and speciation. *Nature* 409: 139-140.
17 Nakamura, M., Yamagish, S. and Nishiumi, I. (2001) Cooperative breeding of the White-headed Vanga *Leptopterus viridis*, an endemic species in Madagascar. *J. Yamashina Inst. Ornithol.* 33: 1-14.
18 Payne, R. B. (1972) Mechanism and control of molt. In *Avian Biology*, Vol. 2. (eds. Farner, D. S. and King, J. R.). pp. 104-155. Academic Press, London.
19 Mizuta, T., Nakamura, M. and Yamagishi, S. (2001) Breeding ecology of Van Dam's Vanga *Xenopirostris damii*, an endemic species in Madagascar. *J. Yamashina Inst. Ornithol.* 33: 15-24.
20 Nakamura, M., Yamagishi, S. and Okamiya, T. (2001) Breeding ecology of the Sickle-billed Vanga *Falculea palliata*, which is endemic to Madagascar. In *Ecological Radiation of Madagascan Endemic Vertebrates* (eds. Yamagishi, S. and Mori, A.). pp. 48-52. Kyoto University.
21 Rakotomanana, H., Nakamura, M. and Yamagishi, S. (2001) Breeding ecology of the endemic Hook-billed Vanga *Vanga curvirostris* in Madagascar. *J. Yamashina Inst. Ornithol.* 33: 25-35
22 Ligon, J. D. (1999) *The Evolution of Avian Breeding Systems*. Oxford University Press, Oxford.
23 Thorstrom, R. and de Roland, Lily-Arison R. (2001) First nest descriptions, nesting biology and food habits for Bernier's Vanga, *Oriolia bernieri*, in Madagascar. *Ostrich* 72: 165-168.
24 Appert, O. (1970) Zur Biologie der Vangawuürger (Vangidae) Südwest-Madagaskars. *Ornithol. Beob.* 67: 101-133.
25 Sato, A., O'hUigin, C., Figueroa, F., Grant, P. R., Grant, B. R., Tichy, H. and Klein, J. (1999) Phylogeney of Darwin's finches as revealed by mtDNA sequences. *Proc Natl Acad Sci USA* 96: 5101-5106.
26 Grant, P. R. (1999) *Ecology and Evolution of Darwin's Finches* (2nd edn). Princeton University Press, Princeton.
27 Putnam, M. S. (1996) Aspects of the breeding biology of Pollen's Vanga (*Xenopirostris polleni*) in southeastern Madagascar. *Auk* 113: 233-236.
28 Eguchi, K., Amano, H. E. and Ymagishi, S. (2001) Roosting, range use and foraging behaiviour of the Sickle-billed Vanga, *Falculea palliata*, in Madagascar. *Ostrich* 72: 127-133.

The evolution of cooperative breeding

Shigeki ASAI and Satoshi YAMAGISHI

Thus far, we have considered the society of the Rufous Vanga from a variety of perspectives. Cooperative breeding has been observed among various birds, and this breeding system is no longer viewed as a new or novel social system among birds. However, only 3% of all bird species display this social system. Despite the fact that cooperative breeding comprises such a minor percentage, research on this topic is disproportionately popular. This is because, from the perspective of evolutionary studies, the existence of "helpers" has been considered a paradox. Some of these studies have been conducted over a long period and in great detail, and collectively, they have presented a detailed documentation of the type of social system in which cooperative breeding occurs. However, the evolutionary development of cooperative breeding and the precise type of process that enabled the development of the present social system remain mysteries. In this chapter, we will introduce several perspectives on the evolution of cooperative breeding, quoting the example of the Rufous Vanga. This chapter will hopefully serve as an overview of the contents of this book.

The benefits of cooperative breeding

Cooperative breeding is a system in which individuals other than the parents care for the chicks in the nest. In several cases, individuals that assist the breeding pair are young birds that have not dispersed but have remained within the natal territory where they were hatched and raised. In order to produce such helping individuals, two processes must be fulfilled. Firstly, individuals must remain within their natal territory and postpone their own breeding (delayed dispersal), and secondly, the individuals that remain must assist the parents in their breeding efforts[1]. The Rufous Vanga is a typical example of this breeding style.

Helpers are known to derive diverse benefits by assisting the pair in its breeding activities. The most commonly cited example is that helpers obtain indirect genetic benefits by helping raise individuals related to themselves. A direct benefit could also be derived—helpers might ultimately gain a territory and a mate. A discussion concerning the benefits that helpers obtain has been provided in Chapter 6. However, in most such species, the benefit that individuals gain by helping is lesser than that which they would obtain by becoming independent and breeding. In the case of the Rufous Vanga, helping has no evident effect (Chapter 6); therefore, even if an indirect benefit exists, it must be extremely marginal. It is considered that individuals would ultimately obtain greater benefits by breeding independently. In general, for inidividual restrained from breeding independently, helping behavior will be the second best tactic. In the case of the Rufous Vanga, we have inferred that helping behavior is the second best tactic (Chapter 7). Consequently, of these two processes leading to cooperative breeding, the delayed dispersal will be brought into focus.

The ecological constraints hypothesis

Two hypotheses were advocated: the "ecological constraints hypothesis"[2] and the "benefits of philopatry hypothesis"[3,4]. The former posits that young birds are unable to participate in breeding because of ecologically constraints, such as the habitat saturation or shortage of mates. In comparison, the latter hypothesis posits that helpers gain benefits by staying in their parent's territory, and therefore, they remain there in order to obtain a territory of good quality and improve their survival rate. Ultimately, two hypotheses proved to be discriminated only by whether emphasis is laid on the cost of departing from parents (dispersal) or the benefit of staying at the natal territory[1,2]. At present, the ecological constraints hypothesis incorporating the benefits of philopatry hypothesis is broadly supported.

The results of several studies have lent support to the ecological constraints hypothesis and have identified shortage of territory, high mortality rate accompanying dispersal, shortage of mates, etc., as the factors that restrain independent breeding. In the case of the Rufous Vanga, it has been suggested that the shortage of mates is the primary factor (Chapter 7). According to the results obtained from the experimental manipulation in breeding units of the Superb Fairy Wren *Malurus cyaneus*, when researchers removed the breeding male from the breeding unit (pair members and helpers), the unit dismantled and a unit was reconfigured in a manner that enabled non-breeding males to hold territories. Moreover, when the breeding pair was experimentally removed and a situation arose wherein there was no female mate although there existed a vacant territory, the non-breeding male in the vicinity did not become independent. However, when the researchers subsequently released the female, the non-breeding male mated with it and took over the territory. Thus, in the Superb Fairy Wren, shortage of mates is a more important restraining factor than territory availability[5]. A number of detailed observational studies and manipulation experiments on various species have provided strong support to the ecological constraints hypothesis. However, the precise reason for the evolution of cooperative breeding in certain species and not others remains unknown. This is primarily because ecological constraints are rather pervasive and their effects are not restricted solely to cooperatively breeding species[6]. Despite the presence of restraints similar to those experienced by cooperatively breeding species, helping behavior has not evolved in non-cooperatively breeding species. Instead, in most such species, non-breeding individuals become floaters or engage in bitter competition in their attempt to obtain a mate. Even among species that always have a helper, some individuals become floaters[2].

Among the Rufous Vangas, males do not solitarily hold a territory and becomes a floater. We believed that this might reflect a lack of available space in which to found a new territory, but this was not the case. When a new territory is established, it often consists of a small space crammed between neighboring territories, and it is difficult to believe that the habitat is saturated. New territories are founded not by solitary males but by both pair members. In a situation wherein only one male is present, it would be difficult for that single individual to engage in territorial conflicts with neighbors; therefore, probably males become independent only after they find a mate (Chapter 6). It has been suggested that ecological constraints that produce sex-specific differences in the mortality rate result in a shortage of mates and that this will lead to delayed dispersal (Chapter 7). Such a situation could have been an important factor in the evolution of cooperative breeding in the case of the Rufous Vanga. However, in several other cooperatively breeding species, both male and female helpers are present, and it is unlikely that mate shortage is a universal factor. Therefore, even if sex difference induced the evolution of cooperative breeding in the case of the Rufous Vanga, it is a condition specific to this species. Thus, it is unable to regard this selection pressure as being common to all cooperatively breeding species.

Comparisons between related species of cooperative breeders and non-cooperative breeders have suggested a correlation between habitat characteristics and cooperative breeding[7]. In the family Vangidae, detailed ecological studies have been conducted only on the Rufous Vanga, and these studies are not yet extensive enough to allow us to undertake comprehensive comparisons. However, cooperative breeding is now known to occur in Chabert's Vanga, the Sickle-billed Vanga, and the White-headed Vanga (Chapter 9). Several species of vangas have been reported as being monogamous (Chapter 9); nevertheless, it can be said that this is a lineage displaying a high incidence of cooperatively breeding species.

Some recent studies have proposed that cooperative breeding is particularly common in environments with warm winters and small annual temperature fluctuation and that foraging ecology is unimportant in this respect[8]. Interestingly, a family-level comparison does not reveal any clear influence of ecological factors on the incidence of cooperative breeding[8]. Accordingly, it is now considered that within a specific lineage, cooperative breeding behavior evolves in species living in ecological conditions that induce cooperation[8].

When considering which groups tend toward a cooperative breeding system, the possibility that a geographical bias could influence the emergence of cooperative breeding is often discussed. Cooperatively breeding species are scarce in the temperate and polar regions and abundant in the tropics. In addition, these species are particularly numerous in Australia[9], where cooperative breeders comprise 10% of all bird species found[10]. Comparative studies among Australian cooperative breeders have revealed that most such species are insectivorous and that they occur in habitats that are stable and show little or no seasonality, and those in which food shortage does not occur[11]. It should be noted, however, that since the species involved belong to lineages whose distribution is largely restricted to Australia, this trend might merely reflect the genealogy of the respective species and not their foraging traits or the environment[12].

Since Madagascar is located at almost the same latitude as the upper part of Australia, it is possible to expect the presence of certain common elements in the evolution of cooperative breeding at these two locations that are separated by the Indian Ocean. The latitude being almost the same, common elements might include a minor seasonal climatic change and an abundance of food throughout the year. If this is so, it would be natural for cooperative breeding to occur at both locations. However, ecological data that would enable a comparison of the quantity of available food and the seasonal variation between Madagascar and Australia is insufficient. The quest for information on the extent to which cooperative breeding has prevailed in avian species in Madagascar has been spearheaded by our study on vangas, and therefore, it might be premature to attempt such a comparison that includes other taxa.

Comparative studies based on the ecological constraints hypothesis have revealed that common elements exist. For example, several cooperatively breeding species occur in environments with less seasonality. However, it can be said that the determinate ecological characteristics responsible for the evolution of cooperative breeding remain unidentified.

The life history hypothesis

Consequently, efforts were focused on identifying life history traits that are associated with cooperative breeding[8,13]. The life history hypothesis emphasizes the importance of life history traits, such as clutch size, dispersal, and longevity, for the evolution of cooperative breeding. With regard to the evolution of birds, life history traits have been regarded as being highly conservative, and in several cases, they are uniform within a single lineage. Therefore, this hypothesis may be capable of satisfactorily explaining why cooperative breeding selectively appears within certain lineages. Recent comparative studies revealed that: (1) the occurrence of cooperative breeding is concentrated within specific families; (2) cooperative breeding is correlated to a low adult mortality rate and a small clutch size; (3) this low adult mortality rate is correlated to philopatry within the territory, distribution at low geographical latitudes, and minor seasonal fluctuations in the environment; and (4) the proportion of cooperatively breeding species in each family is correlated to the typical annual mortality rate of that family[14]. The last factor indicates that low mortality rate itself is not a consequence of cooperative breeding[14], and on the whole, cooperative breeding is more likely to occur within specific lineages.

In the case of the Rufous Vanga, the mortality rate of adult birds is approximately 20%, which is low; clutch size is four on an average, which is small; and philopatry to the natal territory is strong, at least among males. Such life history traits are common to other cooperatively breeding birds; hence, we can suppose that the Rufous Vanga originally possessed traits that facilitated the evolution of cooperative breeding.

The life history hypothesis predicts that cooperative breeding will evolve only in lineages possessing appropriate life history traits, whereas the ecological constraints hypothesis implies that cooperative breeding will occur in any species placed in an appropriate environment[19]. However, as is evident from the results of phylogenetic analyses, cooperative breeding is not necessarily a conservative trait. Further, there are examples of the genus in which non-cooperatively breeding species evolved after the evolution of cooperative breeding[15]. Typically, among cooperatively breeding species, a high proportion of breeding pairs breed monogamously. Moreover, among cooperatively breeding species, there exists a fairly extensive variation in social system. This implies that the social system of cooperatively breeding species is not static. Moreover, these features also occur in non-cooperatively breeding species. According to the life history hypothesis, if a species has an appropriate trait, in other words, if it belongs to a lineage wherein cooperative breeding is more likely to occur, then we can expect cooperative breeding to occur if any species are placed in a suitable environment. An important consideration is whether the converse situation is also feasible—will not evolution of cooperative breeding occur in non-cooperatively breeding species that possess life history traits inappropriate to cooperative breeding? However, even though the Barn Swallows belong to a lineage of non-cooperative breeders, helping behavior has occasionally been observed among them[16]; thus, the difference between cooperatively breeding lineages and non-cooperatively breeding ones is not very distinct. In addition, there exist species that constitute exceptions to both these hypotheses. For example, the Long-tailed Tit undertakes cooperative breeding despite a low adult mortality rate and a habitat that is not saturated[10,17]. Conversely, although they share several traits with cooperatively breeding species, such as high adult survival rate, late onset of breeding, etc., most sea birds do not undertake cooperative breeding[10]. It would seem unrealistic to attempt an explanation of every case in terms of a single model. Indeed, it is a fact that depending on the species, there exist several exceptions. Accordingly, one viewpoint considers it improper that the possession of appropriate life history traits are regarded as a prerequisite for the occurrence of cooperative breeding, and ecological constraints are regarded as the causative factor in the occurrence of cooperative breeding behavior[10]. Perhaps it would be most valid to consider that each case of cooperative breeding has evolved in a unique manner as a result of different natural selection pressures.

The future course of the study on the Rufous Vanga

What insight could studies on the Rufous Vanga contribute to the above discussion? First, regarding the issue of whether the family Vangidae is a lineage prone to cooperative breeding or not, we can state that the family does represent one such lineage (Chapters 8 and 9). However, since the family Vangidae is endemic to Madagascar, its members may not possess life history traits common to cooperatively breeding species; rather, they may only possess ecological traits unique to Madagascar. With respect to the phylogenetic relationships within a family, the lineage does not necessarily reflect the incidence of cooperative society (Chapters 8 and 9). Therefore, it is worth undertaking a comparative study of a variety of social systems within the family Vangidae to ascertain whether the variations are a result of differences in life history traits or are due to differences in ecological characteristics. To achieve this goal, it will be necessary to perform continuous and detailed research, particularly on those vanga species whose ecology is scarcely known.

The phylogenetic analysis revealed the Rufous Vanga and the Helmet Vanga to be very closely related (Chapter 8); yet, the former species is a cooperative breeder while the latter is monogamous (Chapter 9). The ecology of the Helmet Vanga has not yet been described in detail. However, such examples lead us to suppose that ecological characteristics play a significant role in causing these differences in breeding system.

Another approach that we could suggest is a comparison among the populations of the Rufous Vanga inhabiting areas having different ecological characteristics. If sex difference in mortality rate causes a surplus of males and, consequently, gives rise to helpers, then we can infer that even within the same Rufous Vanga species, cooperative breeding will not occur in populations living in an environment where there is no such difference in mortality rate. Comparisons between the society of the Rufous Vanga in rainforests and in dry forest habitats will be particularly informative in this regard.

The study introduced in this book is an important step toward a satisfactory explanation of the evolution of cooperative breeding. As in other studies of cooperatively breeding birds, we were able to achieve certain targets in our research on the cooperative breeding of the Rufous Vanga, notably with respect to how cooperative breeding is maintained in this species. However, further studies on diverse themes are needed to advance our understanding of evolution beyond the current level.

References

1 Emlen, S.T. (1991) Evolution of cooperative breeding in birds and mammals. In *Behavioural Ecology* (eds. Krebs, J.R. and Davies, N.B.), pp. 305-337, Blackwell Scientific, Oxford.

2 Koenig, W.D., Pitelka, F.A., Carmen, W.J., Mumme, R.L. and Stanback, M.T. (1992) The evolution of delayed dispersal in cooperative breeders. *Quarterly Review of Biology* 67: 111-150.

3 Stacey, P.B. and Ligon, J.D. (1987) Territory quality and dispersal options in the acorn woodpecker, and a challenge to the habitat-saturation model of cooperative breeding. *American Naturalist* 130: 654-676.

4 Stacey, P.B. and Ligon, J.D. (1991) The benefits-of-philopatry hypothesis for the evolution of cooperative breeding: variation in territory quality and group size effects. *American Naturalist* 137: 831-846.

5 Pruett-Jones, S.G. and Lewis, M.J. (1990) Sex ratio and habitat limitation promote delayed dispersal in superb fairy wrens. *Nature* 348: 541-542.

6 Smith, J.N.M. (1990) Summary. In: *Cooperative Breeding in Birds: Long-term Studies of Ecology and Behaviour* (eds. Stacey, P.B. and Koenig, W.D.), pp. 593-611, Cambridge University Press, Cambridge.
7 Zack, S. and Ligon, J.D. (1985) Cooperative breeding in *Lanius* shrikes. I. Habitat and demography of two sympatric species. *Auk* 102: 754-765.
8 Arnold, K.E. and Owens, I.P.F. (1999) Cooperative breeding in birds: the role of ecology. *Behavioural Ecology* 10: 465-471.
9 Brown, J.L. (1987) *Helping and Communal Breeding in Birds*. Princeton University Press, Princeton.
10 Hatchwell, B.J. and Komdeur, J.(2000) Ecological constraints, life history traits and the evolution of cooperative breeding. *Animal Behaviour* 59: 1079-1086.
11 Ford, H.A., Bell, H., Nias, R. and Noske, R. (1988) The relationship between ecology and the incidence of cooperative breeding in Australian birds. *Behavioral Ecology and Sociobiology* 22: 239-249.
12 Cockburn, A.(1991) *An Introduction to Evolutionary Ecology*. Blackwell Scientific, Oxford.
13 Rowley, I. and Russell, E.M. (1990) Splendid fairy-wrens: demonstrating the importance of longevity. In: *Cooperative Breeding in Birds: Long-term Studies of Ecology and Behaviour* (eds. Stacey, P.B. and Koenig, W.D.), pp.1-30, Cambridge University Press, Cambridge.
14 Arnold, K.E. and Owens, I. P. F. (1998) Cooperative breeding in birds: a comparative test of the life history hypothesis. *Proceedings of the Royal Society of London Series B* 265: 739-745.
15 Peterson, A.T. and Burt, D.B. (1992) Phylogenetic history of social evolution and habitat use in the *Aphelocoma* jays. *Animal Behaviour* 44: 859-866.
16 Myers, G. R. and Waller, D. W. (1977) Helpers at the nest in barn swallows. Auk 94: 596.
17 Ueno, Y. and Sato, H. (2001) Pair and flock-formation of long-tailed tits *Aegithalos caudatus* in a coastal area of Hiroshima prefecture. *Japanese Journal of Ornithology* 50: 71-84. (in Japanese)

Epilogue

The field site of this study was Madagascar. Animal lovers would long to visit this island at least once because most of the animals found here are endemic species. Moreover, this island is a biological treasure trove where new species of vertebrates are discovered almost every year. Even in our research field, Ampijoroa, a new species of lizard was discovered and described by one of our team members, Dr Tsutomu Hikida (Zoological Laboratory, Graduate School, Kyoto University). The lizard has a strange appearance-the eyes and hind legs are degenerated although the forelegs still remain (Fig 1). In the skink (lizard) family, there are species wherein the forelegs and hind legs are degenerated, but it is highly unusual for only the hind legs to be missing. We can infer that this lizard probably lives under the soil and feeds on earthworms and other invertebrates, but for what purpose are the other forelegs used? Dr Hikida used my name in the nomenclature of this unique creature. This is the new genus and species, *Sirenoscincus yamgishii*[1]. As a research representative of the project, I consider this to be a great honor.

Although I have strayed from the theme, I want to briefly state that the main character of this book, the Rufous Vanga *Schetba rufa*, is one of the species endemic to Madagascar. The research objective was to elucidate the "natural history of the Rufous Vanga" from a comprehensive perspective, including molecular biology, genetics, physiology, morphology, taxonomy, phylogeny, population ecology, community ecology, animal ethology, animal sociology, evolutionary biology, etc. The objective was finally realized with the publication of the present volume, which appeared just before my retirement from Kyoto University. My aim was to ensure that this book would be an exciting and enjoyable read that would encourage young readers to study birds. I would be happy if, after having read this book, they begin to think, "Studying birds seems interesting. Perhaps I too would like to try my hand at it."

This book consists of three parts. Part I describes "why this research began," "where it was conducted," "how our team approached the research on the Rufous Vanga," and "the nature of the Rufous Vanga and the type of

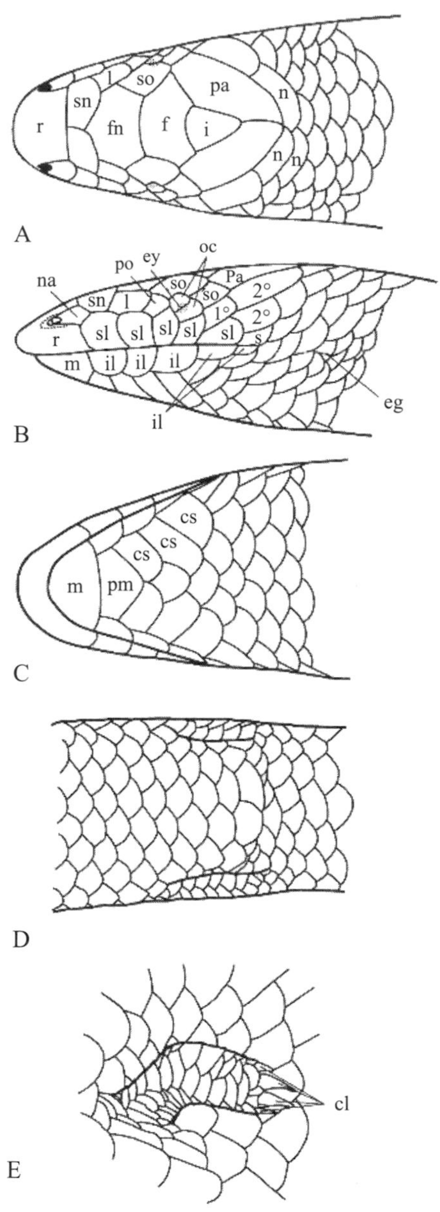

Fig. 1. Dorsal (A), lateral (B) and ventral (C) views of head, ventral view of cloacal region (D), and left forelimb (E) of the holotype of *Sirenoscincus yamagishii* sp. nov. (KUZ R50922). Abbreviations are: 1°, primary temporal; 2°, secondary temporal; cl, claw, cs, chinshield; eg, ear groove; ey, eye; f, frontal; fn, frontonasal; i, interparietal; il, inflalabial; l, loreal; m, mental; n, nuchal; na, nasal; pa, parietal; pm, postmental; po, preocular; oc, ocular; r, rostral; sl, supralabial; sn: supranasal; so, supraocular. (from Sakata and Hikida 2003)

relationship it shares with other birds." Part II is the main segment of the book. Although it includes a description of a rather scientific nature, this is inevitable as we have attempted a demonstrative explanation, making the best use of figures and tables. This part delves into topics such as "ecology and breeding of the Rufous Vanga," "the role of helpers among the Rufous Vangas," and "the reason for the occurrence of helpers." In Part III, "the evolutionary route of the Rufous Vanga in Madagascar" is interwoven with considerations regarding its "morphology, behavior, and social evolution."

I meticulously edited the manuscripts written by each author in order to ensure that each chapter fell into place and that consistency was maintained throughout the book. In the process, some portions of each author's descriptions were placed in other chapters. I also added detailed explanations wherever required. I would like to use this opportunity to express my gratitude to all the authors for granting me the liberty of making such modifications.

I am of the opinion that there are two periods in field research–the "hunting-and-gathering period" and the "agricultural period." The former is a time of mobility and exploration, when we hunt for the research theme and site. I personally prefer this period; therefore, I suffer from the bad habit of oscillating between different themes when the routine work begins. Dr Kazuhiro Eguchi adequately compensated for this weakness. This project involved long-term, detailed, and comprehensive research; therefore, a number of people participated in it. I would like to express my gratitude to these people by citing their names: Hitoha Amano, Isami Ikeuchi, Shingo Fukushima, Masami Hasegawa, Tsutomu Hikida, Masanobu Hotta, Fuminori Iwasaki, Hiromi Kohuji, Mioko Kohno, Tomohisa Masuda, Akira Mori, Hiroyuki Morioka, Hisashi Nagata, Takayoshi Okamiya, Chiemi Saitoh, Yukio Takeda, Masayo Tanimura, Eiichirou Urano, and Norio Yamamura.

Over the period of a decade, several people traveled to the distant island. However, the self discipline and temperate attitude of every single member enabled us to devote ourselves to research without any serious accident.

Collaborators who supported experiments in Japan are Michio Imafuku, Masumi Miyazaki-Kishimoto, Isao Nishiumi, and Chikashi Shimoda. The late Nasolo Rakotoarison, Hajanirina Rakotomanana, Felix Rakotondraparany, Hanitra Rakotonindrainy, Julien Ramanampamonjy, Herilala Randriamahazo, Voara Randrianasolo, and Albert Randrianjafy (names are listed alphabetically and titles are omitted) participated in the collaborative research in Madagascar. We would not have completed our research without the effective cooperation of all these people.

Next, I would like to draw the reader's attention to the ages of the authors. Drs. Honda, Hino, Eguchi, and I were born in 1968, 1959, 1949, and 1939, respectively. As is evident, the participants of this project belong to a range of generations spanning three decades (if we include members in graduate school, this range would extend to nearly four decades). I believe this is a meaningful fact in light of the development of field ornithology and its continued study by future generations.

Research findings must be published as scientific papers. Therefore, as is evident from the references provided in this book, it is a matter of pride that the team members of this project actively published their papers in leading international journals: one each in the Journal of Animal Ecology, the Journal of Molecular Evolution, the Journal of Avian Biology, the Journal of Ethology, and Ornithological Science; six in Ibis; three in Ostrich; and five in the Journal of Yamashina Ornithological Institute. The reader is invited to refer to these papers for further detailed information.

The original articles of this book were published in 2002, titled "Aka-oohashimozu no shakai" in Japanese at Kyoto University Press. This publication was partly accomplished through the advice and efforts of Dr. Andrew Rossiter who acted as translator.

I am extremely grateful to the International Science Bureau, Monbusho (at the time), from whom we received a grant; the Japanese Embassy in Madagascar; the members of Moritani Shokai and Maruha Co. for helping us in Madagascar; Kaori Houki from my laboratory for modifying the figures and tables; Kyoto University Kyoiku Kenkyu Shinko Zaidan for granting us the publication funds for this book; and the staff of Kyoto University Press, particularly Tetsuya Suzuki, for printing this book. The publication of this volume has been supported by a Grant-in-Aid for Publication of Scientific Research Results by the Japan Society for the Promotion of Science (No. 156005 Project head, Satoshi Yamagishi).

Lastly, I would like to thank the entire staff of the Department of Zoology, Graduate School of Science, Kyoto University. Although for a brief period, I spent a comfortable and heart-warming time with all of you. On this, the occasion of my retirement, I would like to dedicate this book to you. It could be termed a "halfway conclusion" of my career as a scientist-halfway because it is "to be continued." I keep pushing myself to make further progress, and it is my sincere wish that you evaluate my work with a critical mind.

Satoshi Yamagishi

On the occasion of retirement from Kyoto University

Reference

1 Sakata S., & Hikida T. (2003) A fossorial lizard with forelimbs only: Description of a new genus and species of Malagasy skink (Reptilia: Squamata: Scincidae). *Current Herpetology*, 22: 9-15.

Index

Subject Index

Organism Index